Harvesting Aquatic Resources

K. P Biswas

2016
Daya Publishing House®
A Division of
Astral International Pvt. Ltd.
New Delhi – 110 002

First Published, 1996
Reprinted, 2016

ISBN: 978-81-7035-158-0 (Hardbound)

Publisher's note:
Every possible effort has been made to ensure that the information contained in this book is accurate at the time of going to press, and the publisher and author cannot accept responsibility for any errors or omissions, however caused. No responsibility for loss or damage occasioned to any person acting, or refraining from action, as a result of the material in this publication can be accepted by the editor, the publisher or the author. The Publisher is not associated with any product or vendor mentioned in the book. The contents of this work are intended to further general scientific research, understanding and discussion only. Readers should consult with a specialist where appropriate.
Every effort has been made to trace the owners of copyright material used in this book, if any. The author and the publisher will be grateful for any omission brought to their notice for acknowledgement in the future editions of the book.

Published by : **Daya Publishing House®**
A Division of
Astral International Pvt. Ltd.
– ISO 9001:2008 Certified Company –
4761-61/23, Ansari Road, Darya Ganj
New Delhi - 110 002
Phone: 011-4354 9197, 2327 8134
E-mail: info@astralint.com
Website: www.astralint.com

Laser Typesetting : **Classic Computer Services**

Printed at : **DP's Impressive Impression, Delhi**

PRINTED IN INDIA

**Dedicated to my wife MANJU
for her constant encouragement
in writing this book**

PREFACE

The original forms of life on the earth were microscopic water plants, called phytoplanktons. Phytoplanktons drift in enormous quantities in the surface waters penetrated by sun–rays and form carbohydrates and other organic materials through photosynthesis, using the solar energy and nutrient salts in the water.

The most important and numerous phytoplanktons are diatoms. They are primary producers in the marine ecosystem. It is said that in the oceans of the world 10 billion tonnes (in dry weight) of phytoplankton are produced in a year.

Phytoplankton together with zooplanktons form valuable foods for juveniles and larvae of fishes. There are many fishes feeding mainly on the phytoplankton even in their adult stage. It is well known that sardine feeds on diatoms and flagellates, sweet fish feeds on diatoms and blue-green algae and milkfish feeds on blue-green algae.

Organisms cannot live in isolation. All animals live on other organisms, while they are in turn, preyed on by some other animals. There are also competition between different animals for common prey. These food relations are most important among living organisms. Animals and plants inhabiting in waters have certain "food chain" relations, which find each animal and plant holding a certain ecological position in the habitat.

However, food chain relations are never uniform. In each individual habitat, there is a unique ecosystem consisting of certain food chains. Even in the same habitat, the patterns of food chains also vary with the seasons, as animals feed on different organisms during different stages of their life history, e.g. fry, yearlings, adults.

The patterns of food chains reflect the structure of reproduction of the given area. Therefore, it is essential to understand the ecosystem in which fishery resource exist for effective exploitation and sustainable yield.

To obtain a greater harvest, it is necessary to go for the activities to enable increased production by stabilizing and facilitating the increased natural reproduction process in the ecosystem in one hand, together with rational exploitation of the living marine resources (fish, prawn, shell fish, sea weeds) on the other hand on a sustainable basis.

All marine animals have evolved to fit the requirements of their respective natural environments, and every species has secured its own ecological niche in order to survive. Fish like herring and sardine that feed on planktons have a high reproductive capacity but their ecological niche is relatively unstable and subject to unpredictable patterns of large, cyclical changes. In contrast, fish higher up on the food chain like tuna and flounder are less productive but have a more secure ecological niche. Although they are not subject to large natural fluctuations, they are more susceptible to mortality caused by fisheries (the catch).

Fishing industry is like one big "integrated circuit" and functions on the basis of a continuing interaction of three basic elements: (1) the reproductive capacity of the natural resources, (2) fishery and (3) the effective use of what is produced; in other words, the catch. It is only when these three elements are in an appropriate balance, then we can expect a fishing industry to develop and grow in a healthy and productive way.

As for fishery, it is a multi-faceted activity that involves fishing grounds, boats and labour and is ultimately dominated and restricted by the fish-catching capabilities of the fishermen, which is referred as fishing technology.

A wide variety of curious gears for herding and catching fish promises to elevate the ancient art of fishing to a technological par. Yet there are kinds of fish which inhabit waters where our present gears cannot work; and which have such habits that our present gears cannot take them. Many people believe that there are large stocks of mid-water fish in the open oceans far out from land, and awaiting exploitation.

The single most important development in fishery since the beginning of recorded history has been the motorization of fishing boats, leading to the "industrialization" of fishery.

By definition, a fishing boat must perform four basic functions. These are: (1) mobility (the capability to travel over a given distance of water), (2) the ability to locate fishing grounds, (3) the capacity to perform fishing operation and (4) the capacity to transport the catch. Of course, over the long history of fishery, mankind developed techniques to harness wind power to propel his boats over long distances and use the power of currents to operate his fishing nets, but none of these techniques could compare with the advantages of motorization of fishing boats. The major operational advantages are:

(a) it ensures a given level of high horse power capability;
(b) long hours of continuous operation; and
(c) devices for mechanizing the fishing operation.

These encouraged the building of bigger fishing boats and made it possible for fishermen to seek out fishing grounds further from shore. Also, it not only encouraged the mechanization of the fishing operation, but also led to the development of new systems of mechanization that economized traditional fishing methods and further led to attempts to develop entirely new fishing methods.

Major ancient fishing gears and methods, together with today's harvesting methods of living aquatic resources have been included in the book for the benefits of those who are concerned with fishing industry. Advancement and adaptation in fishing technology and its future prospects and dreams have also been outlined in the edition.

K. P. Biswas

CONTENTS

CONTENTS

INTRODUCTION

The total area of the oceans and adjacent seas is 361 million sq. km., account for 70 per cent of the total surface of the globe; $510 \times 10^6 km^2$. A gentle slope extends out from most shorelines to 100–200 m., known as continental shelf, common to almost all the sea areas in the world. The continental shelf is thought to have been a part of the land at the time of glacial age. From the continental shelf, the sea bottom slopes down rather steeply, the continental slope. Reaching a depth of about 3,000 m., the sea bottom resumes a gentle slope, until it reaches a depth of about 6,000 m., which is called the "abyss" or deep sea floor. Trenches of 6,000–10,000 m. deep are located at points on the deep sea floors. Tonga trench and Porto Rico trench are located at depth of 7,632 m., and 7,900 m., respectively. The Philippine trench (10, 266m.) is considered to be the deepest area of the oceans of the world.

Although, the seas cover 72 per cent of the globe, they yield only one per cent of our food. A remarkably little is known of the dark ocean depths that literally lap our modern civilization. The capture of living Coelacanths, supposedly extinct for some fifty million years, is only one indication, though a particuarly striking one, of the limits of our knowledge about the living resources of the sea.

Biologists and oceanographers call the waters above continental shelves "coastal waters" differentiating them from "high seas waters". Although the coastal waters constitute, in terms of area, only 9.9 per cent of all the oceans, they are very rich in biota, with the productivity of the living resources being estimated at approximately two times that of the high sea waters.

Ecological zonation of the oceanic depths are the main features for existence of life in seas. The organism of even the deepest trenches are dependent on the biological conditions of the overlying water masses, to the very surface of the ocean. At the surface, light is the dominant factor in the epipelagic zone. The essential ecological feature of the photic zone is that the animals live in the same environment, with the primary food source, the phytoplankton, whereas in the lower layers there is no primary production by plants. Although the endemic species of the epipelagic zone either do not migrate or perform only limited vertical migrations, there are many animals, which invade the epipelagic zone from the deeper layers during the night or pass their early developmental stages in the photic zone. In either case, the abundance of food is the primary factor.

The epipelagic zone is a very thin layer. In eastern parts of the oceans, with upwelling and high productivity, it may be less than 100 m. thick, and in clear subtropical areas it is 200 m. thick.

All organisms living below the photosynthetic zone depend upon the food (energy stored in organic matter, as such energy can be utilised by heterotrophic bacteria, when they assimilate plant and animal remains or dissovled organic matter) that falls from the uppermost zone. According to ZoBell (1954), bacteria oxidize approximately 60-70 per cent of organic carbon to carbon dioxide, whereas the remaining 30-40 per cent of the organic carbon is converted into bacterial cell substance (particulate protoplasm), consisting primarily of proteins and lipids that ultimately serve as an important source of food for the deep-sea animals. Marine animals, like, protozoans, bottom grazers, filter feeders, and mud eaters, ingest and digest bacteria. Remnants of land plants, such as, branches, leaves, fruits or marine algae from the coastal belt have been seen in many hauls from deep water, even at the bottom of the Philippine Trench (Agassiz, 1888; Mortensen, 1938).

The food supply in the abyssal and hadal zones, on the whole, is not plentiful, but no more scare than in any other region of the ocean below the layers that are in fairly close contact with the photosynthetic zone. The actual depth is of minor importance in comparison with the nearness of the land and the productivity at the surface. The amount of dissolved organic matter represents a fairly uniform supply, so that no abyssal desert, without animal life, is found because of lack of food, where other conditions of life are fulfilled.

The division of the hydrosphere into a warm troposphere and a cold stratosphere, is the most pronounced physical division in the oceanic water masses– whereas temperature is an important factor governing the horizontal distribution of the animals of the epipelagic zone, in the lower zones of the sea, temperature is more important in influencing the vertical distribution of animals. Some mesopelagic animals, are very rarely or never caught near the surface due to higher temperatures of the upper layers. Downward, however, the limit set by temperature is much more pronounced. There is a fairly pronounced discontinuity layer between these regions in most tropical and subtropical seas. The most useful ecological boundary is marked by the 10°C isotherm; also the surface isotherm at the limits of the polar regions.

Photosynthetic zone, with the primary food, the phytoplankton, extend downward as far as the sunlight penetrates during the day with temperature more than 10°C (thermosphere). A zone of dim light leads down into the always dark mesopelagic zone, from which many species rise to higher levels during the night. For food the mesopelagic animals, as those of deeper zones, depend on the epipelagic zone. Downward the discontinuity layer, separating the thermosphere from the psychrosphere leads into the bathypelagic zone at temperature below 10°C, where food gradually becomes scarce; food becomes increasingly scarcer in the abyssopelagic zone from temperature about 4°C and lower. Close to the bottom an accumulation of organic debris from above starts a secondary production by heterotrophic organisms, serving as the first link of a new chain in the benthic abyssal zone.

The original forms of life on the earth were microscopic water plants, phytoplanktons. Phytoplanktons drift in enormous quantities in the surface waters penetrated by sun–rays and form carbohydrates and other organic materials through photosynthesis, using the solar energy and nutrient salts in the water. The most important and numerous phytoplanktons are diatoms, the primary producers in the marine ecosystem. Coastal waters often have upwellings or vortexes, which cause vertical mixing or agitation of sea water, providing biological environments, facilitating the propagation and growth of planktons. The nutrient salts in inflowing land waters also facilitate propagation of phytoplanktons and zooplanktons in the coastal waters, on which fishes and other marine animals feed.

All forms of life have undergone a long evolutionary process to reach their present forms and places in the ecosystem. Each species has its own habitat suitable to its life cycle. Except for plants, which are capable of forming nutritive substance for themselves, all animals live on other organisms, while they are, in turn, preyed on by some other animals.

Competition also exists between different animals for common preys. Animals and plants inhabiting in an area have certain "food chain", relations which find each animal and plant holding a certain ecological position in the habitat. The patterns of food chains reflect the structure of productivity of the area.

Aquatic plants grow as they produce carbonic compounds by photosynthesis using solar energy and nutritive salts dissolved in the water. Simplest form of marine plants are phytoplankton, such as, diatoms. It is said that in the oceans of the world ten billion tonnes (dry weight) of phytoplankton are produced in a year. Phytoplanktons together with zooplanktons are valuable foods for juveniles and larvae of fishes. There are many fishes feeding mainly on the phytoplankton even in their adult stage. Also many snails live on the aquatic weeds. Similar to that of terrestrial plants, in hydrosphere, depending on the region, depth, and conditions of the tidal currents various aquatic plants adapt to each habitat grow abundantly. The existence of aquatic weed is controlled by the quantity of radiant energy from the sun. The range of light penetration varies with the area, and it has been estimated to be about 200m. depth. Almost all species of aquatic weed, therefore, grow in water shallower than 50 m., and especially shallower than 20 m. Only a few species and small number of seaweeds can grow on the seafloor at 100-200 m. depth. Useful seaweeds, such as, green algae grow in shallow water; red algae in the deep water; and brown algae in intermediate depth between the first two.

In recent years, "sea vegetables" has come to be used as esteemed "natural foods" or "medical foods", rich in nutritive substances (Vitamins, minerals etc.) necessary for good health. Some species of seaweeds contain many polysaccharides and proteins as well as vitamins and minerals, useful for man and livestock after extracting these in digestable form; besides using them (seaweeds) as feed and fertilizer.

Thus, the major portion of the world population, sought sources of protein for their diet in the living aquatic resources, such as, fish, shellfishes, crustaceans and even seaweeds of the abundant sea, the "gift of the sea".

Until the development of chemical fertilizers in recent years, mass-catch fish, such as, sardine, saury, and herring were an important source of fertilizer for agriculture.

The commercial value of aquatic products does not remain constant, but changes with the demands of the times. These changes, however, are not limited to the economic laws of supply and demand; but with the improvement in harvesting technology lead to changes in supply structure, that in turn cause changes in product value.

One of the main characteristics of living aquatic resources as an industry is the fluctuations in production. Some species are particularly susceptible to overexploitation; while in other cases large-scale changes in the sea environment can also cause drastic changes in resources. With fishery and marine food processing developed into large industries, as they are today, such changes in aquatic resources can have a major effect on human society.

In this context, the harvest of living aquatic resources, on a sustainable basis, is of prime importance, the different facets of which have been discussed in this communication.

1

BIOTIC COMMUNITIES

Utilization of energy from bio-mass resources have been receiving wide attention today. "Biomass" is a concept in ecology, and it indicates the physical quantity of "solar energy fixed as organic matter by chloroplasts in plants". The amount of energy coming to the earth from the sun reaches seven hundred million times of one trillion kilocalories, but only a thousandth is stored in form of starch by photosynthesis of plants. Practical use of biomass means, the extraction of some substances, such as, methane and ethanol from waste portions like stems and leaves of the plants other than those portion used as food.

The use of aquatic plants (marine) as a new energy source to obtain fuel, such as methane and other by products, like alginic acid, fertilizer, livestock feed, food and medicines have been commercially operated in many developed countries of the world.

Important conditions for growth of aquatic weeds, next to light are water temperature, salinity, waves, emergence during low tide, and substratum for attachment. Seaweeds unlike terrestrial grasses, absorb nutrients through the whole leaf-like body, and the holdfast of seaweed corresponding to the root of terrestrial plant functions only to attach the leaf-like body to the substratum for growth. Except planktonic unicellular algae, the habitat of seaweeds is generally limited to the place where fixed substratum for attachment, such as, rock bed, reef, boulder, gravel and concrete block is available.

Seaweed beds are the infant cradles of the sea. These beds where seaweeds or seagrasses grow densely over a sufficiently large area of the sea bottom often play a vital role as spawning and nurturing grounds or habitats for aquatic organisms. Seaweed, beds can be of number of types. Zostera beds made up mainly of seagrass; Sargssum beds made up mainly of gulfweed; laver beds of *Eisenia bicyclis* besides Kelp beds. Zostera is a flowering plant that proliferates in the waters of inland sea and bays with gentle currents, sand and mud bottoms at depths up to 5-6 metres. Zostera beds with their bright green grasses give a special appearance to shallow sea areas and serve as the nurturing grounds for numerous fishes, such as, black sea bream, sea bass, black rockfish and swimming crab during their larval stages.

Some fish like black sea bream and black rockfish continue to make the Zostera bed as part of their habitat even after reaching maturity. For such fishes, these beds serve both as a feeding ground rich in small marine animals that live on the grass surface or species like sand worm and small shrimps, as well as a place to hide themselves from predators. Cuttle fish and certain species of fish spawn by attaching their eggs to the surface of Zostera grasses.

Animal Communities

Among innumerable aquatic animals, fish, crustaceans, shell fishes and mammals are of particular interest with regards to their usefulness to mankind. For judicial and sustainable harvest of these aquatic resources it is necessary to understand their mode of life and the patterns of their movement, so as to devise introduction of effective kind of artificial stimulation.

Aquatic animals move seasonally in search of proper water temperature, food and breeding spots. The knowledge of this behaviour, migration of aquatic animals, are extremely important for the fishermen to know. Tuna migrate to areas where the water temperature is 14–15°C, therefore, they will not be found outside such areas. Salmon always return to their home area for breeding. At this time salmon can easily be taken in shallow rivers without extensive equipment. Many fishing methods make use of these habits and migrations of aquatic animals.

Fish search for the appropriate temperature and salinity for spawning. Consequently in summer they go to colder north areas and in winter they migrate to warmer south seas. Salmon leave the sea and enter in rivers for breeding; while eel leave river and enter the sea for spawning. Eggs which hatch in shallow coastal areas migrate gradually to deeper sea as they mature, usually returning to their original spawning area. Fish migrate looking for more food at same depths; while deep sea fishes remain in deeper area in the daytime and come up to the surface at night in search of food.

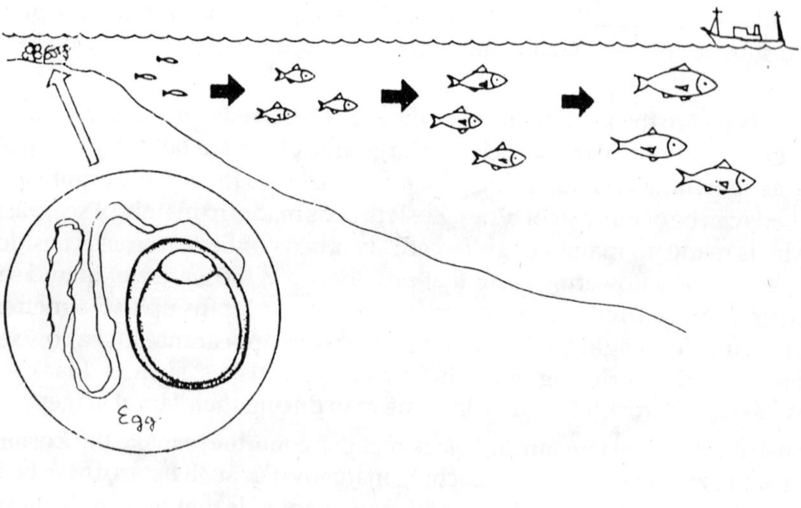

Fig. 1: Maturation migration: Horse mackerel

Fig. 2: Breeding migration of Salmon and Eel

Fish community can be divided into cold water, warm water and tropical water species. Vigorously swimming fish search for waters of the proper temperature; while weak ones swim on ocean currents. Fish are abundantly distributed in areas where ocean currents meet. Some fish breed only in a fixed area, but they are not numerous. In tropical and warm seas, fish found in deep areas are the same as those found in sub-arctic seas. Same fish, again, are sometimes found near the surface. Even so, those fish which seldom go into the deep seas pass the edge of continental shelf (200 m. or shallower). There are more than 400 species of fish that are known to live at depths greater than 2000 m. and many fish are known to live at depths up to 200 m. Commercially important fishes, however, live in relatively shallow of 200 m. depth or less. Most of them are coastal fish and inhabit within 10-30 metres of the sea surface.

With these knowledge of fish habitat and migration, different harvesting methods have been evolved. Given below the informations about the habitat of some of the commercially important fish, shell fish and crustaceans.

Fish Living in Rivers & Lakes

1. Bull head — 15 cm. body length, lives in clean rivers.
2. Catfish — 50 cm. body length, with no scales but long barbels lives in rivers and ponds.

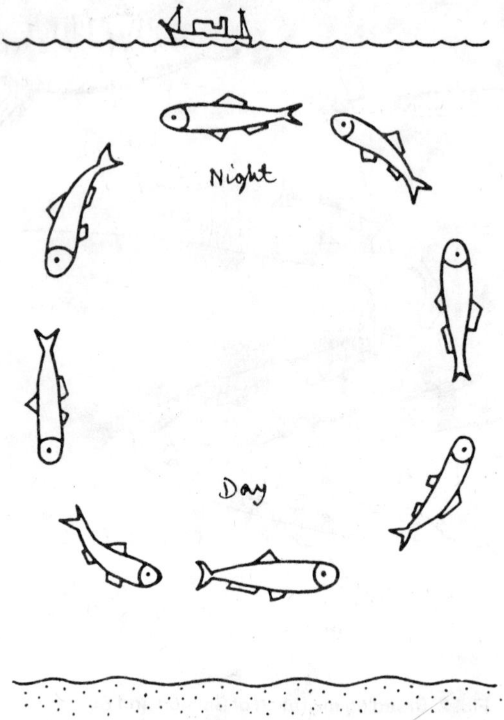

Fig. 3: Feeding migration

3. Minnow	—	20-30 cm. body length with large mouth. Lives near surface.
4. Roach	—	20 cm. body length, with variations in body colour. Lives in sandy bottoms and lay eggs on backs of shell.
5. Perch	—	5-10 cm. body length; lay eggs on backs of shellfish.
6. Smelt	—	15 cm. body length; lives in mountain lakes also.
7. Shiner	—	5-13 cm. body length; prefers area where the current is gentle.
8. Dace	—	30 cm. body length; an important food fish.
9. Bull trout	—	40 cm. body length; a typical sport fish lives in upper streams of mountain.
10. Rain bow trout	—	30-40 cm. body length; lives in cold mountain streams.

11.	Crucian carp	—	20 cm. body length, matures rapidly.
12.	Loach	—	18 cm. body length. Lives in muddy bottom of pond and rivers.
13.	Carp	—	60 cm. body length; most important freshwater fish.
14.	Sweet smelt	—	30 cm. body length; lives in clean rivers.
15.	Eel	—	60-100 cm. body length; lives in river and ponds.
16.	Seabass	—	100 cm. body length; ascend in river in spring and summer.
17.	Grey mullet	—	30-40 cm. body length; lives in estuaries.

Fish Living Off Shore at a Depth Below 20 Metres

1.	Dorado	—	180 cm. body length. Found in warm seas and has a habit of gathering around driftwood or ship bottom.
2.	Black rock-fish	—	20-30 cm. body length.
3.	Octopus	—	75 cm. body length. Found in shallow seas.
4.	Sea slug	—	20 cm. body length, with soft body and bones.
5.	Ice fish	—	10 cm. body length; semi-transparent.
6.	Prawn		22 cm. body length; lives in sand bottoms at depths around 20 m.
7.	Abalone	—	12 cm. body length; lives near shore, active at night.
8.	Scorpion fish	—	30 cm. body length; caught by angling.
9.	Scarus	—	25 cm. body length; at night burrows on sandy bottom and sleeps.
10.	Calm	—	4-6 cm. bivalve; lives in bays.
11.	Sillago	—	30 cm. body length; lives in sandy bottoms of shallow seas.
12.	Sea bream	—	30 cm. body length; found in coastal waters.
13.	Flathead	—	50 cm. body length; lives in sandy bottom.
14.	Goby	—	20 cm. body length; found in coastal areas in gentle currents.
15.	File fish	—	15-20 cm. body length; lives in coastal waters.
16.	Mantis shrimp	—	15 cm. body length; lives in sand and mud.
17.	Sea urchin	—	6 cm.; found in shallow waters near shores.

Fish Living at a Depth of 20-200 Metres

1.	Shrimp	—	5 cm. body length; transparent, when alive; comes to the surface at night.

2. Herring — 35 cm. body length.
3. Mackerel pike — 40 cm. body length; migrates seasonally; lives in 14-15°C water temperature.
4. Lockington — 40 cm. body length; lives near shores.
5. Flounder — 80 cm. body length; found in sand and mud areas.
6. Red seabream — 120 cm. body length.
7. Squid — 30-35 cm. body length; lives in outer seas; stays at a depth of 10 m. during day but come to the surface at night.
8. Hairtail — 150 cm. body length; Warm water in habitants; comes to the surface in morning and evening.
9. King Crab — 25 cm. shell width; lives in cold water.
10. Sardine — 25 cm. body length; inhabits near the surface.
11. Black porgy — 45 cm. body length, found inside bays on muddy bottoms.
12. Sciana — 100 cm. body length; found near shore.
13. Yellowtail — 100 cm. body length; lives in warm seas.
14. Tuna — 300 cm. body length.
15. Frigate mackerel — 40 cm. body length.
16. Striped pigfish — 40 cm. body length; prefers weedy shore, comes to the surface at night.
17. Gizzard shad — 25 cm. body length, found in bays.
18. Sting ray — 100 cm. body length; found near shores.
19. Lobster — 36 cm. body length; lives at depths up to 50 m.
20. Flying fish — 35 cm. body length, can skim along the surface for a distance of 30-50 m.
21. Horse mackerel — 40 cm. body length; can swim long distance.
21a. Launce — 20 cm. body length; likes sandy bottoms along shores.
22. Black rock fish — 30 cm . body length; caught in angling.
23. Tiger shrimp — 27 cm. body length; lives in sandy bottoms at depth up to 60 m.
24. Butterfly fish — 18 cm. body length; lives in warm seas.
25. Cod — 120 cm. body length.
26. Sea eel — 100 cm. body length, tropical.
27. Lizard fish — 50 cm. body length; buries body in sandy sea bottoms.
28. Salmon — 90 cm. body length, ascend rivers for spawning.
29. Mackerel — 45 cm. body length.
30. Skip jack — 90 cm. body length.

31.	Rock bass	—	90 cm. body length.
32.	Sword fish	—	200 cm. body length; lives in warm deep seas.
33.	Blow fish	—	30 cm. body length; lives along shores.
34.	Snake mackerel	—	40 cm. body length; move in large schools near the surface.
35.	Albacore	—	150 cm. body length.
36.	Frogfish	—	100 cm. body length; lives on shores.
37.	Cuttlefish	—	15 cm. body length; lay eggs on the shore.
38.	Scallop	—	15 cm. lives in pebbly sea bottoms at depths of 30 m.

Fish Living at a Depth of Over 200 Metres

1.	Sand fish	—	15 cm. body length.
2.	Flounder	—	40 cm. body length.
3.	Fat bass	—	200 cm. body length; lives near shore.
4.	Alaska pollock	—	60 cm. body length.
5.	Big eye	—	60 cm. body length.
6.	Perch	—	100 cm. body length.
7.	Red coral	—	Grows in large branches, available in waters of 18.5°C or higher temperatures.

Elements of Harvesting Methods

Techniques of catching aquatic animals began as a process of outsmarting the natural prey, as in hunting on land. All of mankind's harvesting technology began by first understanding the life mode and the patterns of movement of the aquatic animals and, from this understanding, devising ways to introduce some effective kind of artificial stimulation. Mankind has always invented tools to make up for his physical limitations, and it is known that even prehistoric man had invented a wide range of harvesting methods and gears, including boats and fish hooks as well as nets and fish weirs. The basic principles on which today's fishing gear and methods are based can all be traced back to these primitive methods.

Fishing is considered to be the earliest form of hunting and the world's oldest industry. Engels maintains that the middle stage of the savage stage began, when man started catching fish for food and using fire.

As a regular occupation, fishing began some 50000 years before our era. At first the most simple and primitive form of fishing was used—catching the fish with bare hands. Those who caught fish at that time very much resembled animals doing the same. The first fishing implements began to appear about 25000 years before the present era. During that period, fishing developed, together with hunting and this is evidenced by the fact that the first fishing implements were equally suitable for use by either, hunter or fisherman. Simple though such implements were, they started the fishing methods in which man began to use some kind of tackle.

Originally, such tackle was extremely primitive. Man in quest of harvesting fish tried to improve it. What he was chiefly after, was such devices as would lengthen his arm during fishing. In this, he followed the hunter who, in order to bring game within his reach, gradually replaced the club with the spear and the harpoon, and later with bow and arrow. In this way the reach of man's arm was multiplied many times. In the modern life of many countries the art and practice of fishing has been divided into two diverse fields; the sport fishing and commercial fishing. While the sports fishermen strives for the catch of capital gain, the commercial fishermen works for mass production. Moreover, when the term fishing is mentioned, it does not only mean the taking of fish; but also relates to the capture of all aquatic animals, such as, crustaceans, molluscs, coelenterates, birds, mammals and even water serpents. All animals even frogs and turtles are legally regarded as fish in many countries.

The techniques of fishing include fishing methods as well as fishing gear. Fishing methods mean the manner or way in which fish may be captured; while the fishing gear is the implement developed for the purpose. But often the gear alone is not sufficient and ancilliary instruments have to be used. A fishing method can be applied by means of various gears, just as a fishing gear can sometimes be used in the application of several methods.

On comparing the fishing methods of various nations, it becomes evident that the fishing techniques have been developed from only a very few basic ideas for capturing fish either singly, or in a mass for the benefit of man.

Evolution of Harvesting Methods

Initially all the ideas have commenced with simple fishing methods with primitive gears. Later more complicated techniques emerge. This development has been hastened and improved by various impulses. One such impulse at least in commercial fisheries, was that it became desirable to change from the catching of single fish to catching fish in bulk. Fishing gears capable of catching those masses, had, therefore to be enlarged and their efficiency improved. That required at one and the same time mechanization of the operation and the designing of more effective gears.

Another impulse, which stimulated further development was the need to abandon the shallow waters and penetrate into greater depths in order to explore new possibilities of bigger supplies of fish. This separation from the shore depends in the first instance on the fishing vessel. Thus parallel with the development of the fishing technique, there occurs the enlargement and improvement of fishing craft from the bamboo raft, still much used in Asia for short time fishing to the factory ship, capable of staying at sea for many months and processing the catch immediately on board.

Further stimulant for developing fishing methods and gears is the desire to progress from the originally guarded or watched fishing gear to the automatic unguarded gear, that is, to close the gear in time to prevent the fish from escaping. Recently the Japanese have set "sono-buoys" in pound nets to carry by wireless to considerable distances the news of the catch, its kind and quantity, by transmitting the underwater noises made by the fish.

The desire to have gears, to be used for several purposes has also influenced the further development of fishing gears. These are gears which may be used in many ways, such as, gill nets, able for drift net as well as for set net fisheries, or trawl nets, which are able to fish near the bottom or surface or midwater as well. Purse seines to be used for big schools, as well as for small shoals without the whole net being set belong to that category. For that reason, so-called combination vessels are used which can operate at least two of the various fishing methods.

In many industrialized countries, with shortage of labour, the mechanization of fishing methods, which does not only refer to mechanization of fishing vessels; but also to fishing techniques, can solve the manpower problems.

Basic Ideas for Development of Harvesting Gears

The development of harvesting methods proceeded step by step from the capture of individual fish to catching them in bulk. Accordingly, fishing methods gradually changed from early techniques to the use of many ingenious and sometimes large and complicated gears for harvesting the riches of the sea. But the basic ideas of fishing have not changed, rather those ideas have transformed the original simple concepts into forms, wherein they could meet the modern demand for greater production under more rationalistic conditions. Nearly all the basic ideas for catching fish have been developed from their original simple form into effective modern fishing practices. The cast spear has been transformed into fired harpoon; the scoop net, operated by hand, has been developed into the trawl, towed by large vessels over the bottom of the sea; the simple hand line with a single hook has been turned into the expensive highly developed catching equipment of the sports fishermen and into the long line equipped with many thousands of hooks used by commercial fishermen.

The importance of the gear varies according to the point of view from which each particular gear is regarded. In relation to fishing yield, the trawls and purse seines take first place. Long lines and gill nets can be of great general importance. In local applications dip nets and some types of traps are useful. With the exception of fishing with pump and using electricity for fishing, no substantial new methods of how to catch fish have been developed in the past hundred years. The highly developed trawls and purse seines are no exception to this. Not only, the basic ideas for fishing methods are limited it is also striking and impressive that only a small number of fundamental building elements have been used in the construction of the most varied fishing gears.

The barbed point is a building element, that is used with spears and similar gear, led eventually to the development of the modern whaling harpoon. The barbed point is used also in angling and even in traps. But the barb is not a specific element for fishing, it is used also in hunting with certain gear that is designed for the same purpose.

Netting or strictly speaking the mesh is a buidling element which fishing has in common with hunting for the manufacture of catching gear. Netting, however, is now used as a building material in almost all fishing gear. The use of net for catching animals has been copied by man from nature. The traps used in both hunting and fishing can be identical. The funnel-shaped non-return device is an element that has been utilised in both hunting and fishing. Funnel shaped valves are used in building of barriers and fish baskets; of small and large fyke nets, of trawl and seine nets as well as stow nets. In hunting such implements are represented by many traps and latticed gates.

Typical of fishing are those building elements, which can become effective only in water. These include the system of floats and sinkers which serve to keep net walls and whole fishing gears upright or floating. The system is used with seine nets and trawls, with surrounding nets and drive-in net, but especially with gill nets and those that capture fish by entangling.

For dragged fishing gears floats and sinkers can have the function, not only of enlarging the aperture, but also of keeping the fishing gear in the correct position, when it is set before the swell of the water during towing brings the fishing gear itself into proper shape. With various fishing gears, floats are used on the surface of the water to keep the gear afloat.

The pressure of swell created during towing, however, can cause the floats to loose their buoyancy and begin to act as brakes or retarders. To minimise the undesirable braking effect of floats, they can be converted into shearing devices. In that case the buoyancy is increased by the shearing power created by towing and the hampering effect of floats, when the gear is being dragged can be reduced or prevented. Simpler shearing devices are often used in the form of boards with the purpose of spreading a gear wider during towing or when it is in current. The use of otter boards in trawling made it possible to relinquish the use of beam for keeping the net open horizontally. Shearing devices do not only keep a gear open in the horizontal direction, they are also used for keeping the net open vertically by shearing like the Kites on the headlines of trawls or as depressors for lines, dredges and beam trawls or mid-water trawls.

A different building element that is used more often for manufacturing fishing gear is that of small pockets for capturing and retaining fish that seek to escape from the gear.

For guiding purposes, vertical stretched lines or lines of stakes arranged at considerable distances are used. In trawl nets, kites floating ahead of the net enlarge the aperture of the net due to their swinging warps or connecting lines. With the bamboo traps of Thailand, the bamboo stakes set alternately in the bottom serve as a guiding device inspite of the intervals between them being large enough to let the fish escape.

Quite different to the guidelines, signal lines are stretched individually in the openings of the fishing gear, the other end is held by the fisherman's hand. As the fish push past into the gear, the fisherman feels the vibrations and can feel the number, species and size of the fish going into the gear. Signal lines are thus the predecessors of our modern fish detecting devices.

Elements of Fishing Methods

Numerous fishing methods, operating today, are based on a few basic principles. It is necessary to consider the necessary elements first, in order to analyse the various fishing methods and equipments. Methods include both frightening and enticing fish. When frightened, fish naturally disperse and as such this method can not be used for any length of time. The method of enticing, however, is found effective to gather fish and is, therefore, a superior method. Between the above two, is the pound net method in which nets are used to capture marine life from the sea. For this reason, the pound net has come into wide use.

Most of these fishing nets are made of woven threads, but even single pole, electric current, sound or light can be employed to form a sort of abstract net. The usage of nets includes the passive method of stationary nets and the more effective method of moving nets.

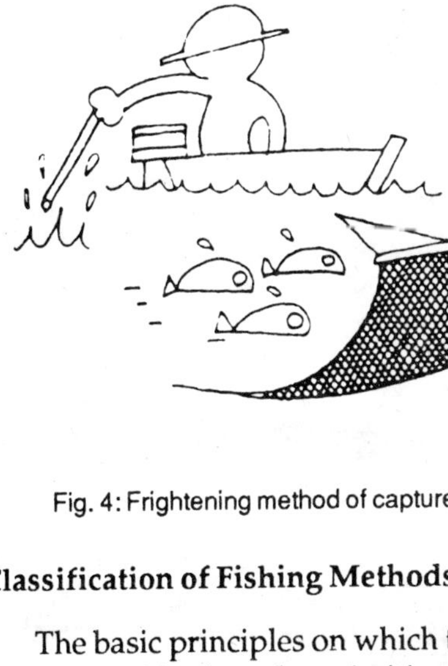

Fig. 4: Frightening method of capture

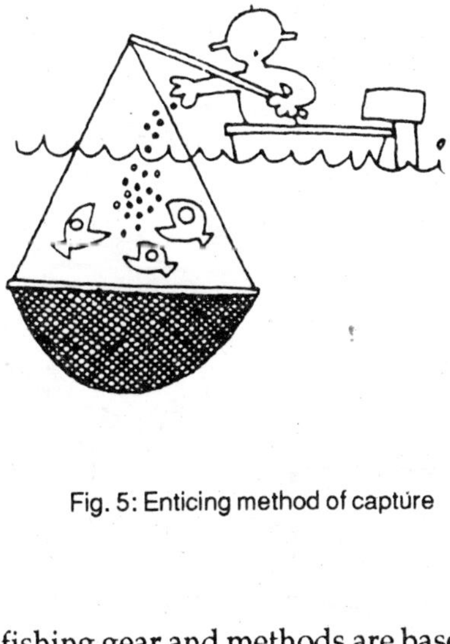

Fig. 5: Enticing method of capture

Classification of Fishing Methods

The basic principles on which today's fishing gear and methods are based can all be traced back to the primitive methods, when man became proficient in one method, he began to improve that method through trial and error. In this process, the introduction of new material often led to big advances in proficiency. In fact, to a large degree, the history of development of fishing technology has been the history of advancement in materials.

There was one more element that played a major role in the development of fishing technology. That element was information. Fishermen have always had to be well informed about the distribution of the fish resources, the timing of their migrations and the seasonal changes in the amount of resources.

In Japan, between 17th and 18th centuries, various types of net fisheries began to flourish, and all of the basic forms of net fishery in use today can be traced back to this period. During 16th to 19th centuries, there was development of specific fishing methods aimed at specific species of fish, and with this, the appearance of certain representative types of fishery in the different regions of the country came into existence.

The biggest advancements, in the development of fishing technologies have come with the advent of mechanization in modern times. The overwhelming power and capability for continuous operation of motors compared to manual labour brought true "industrializatioin" to fishery, and along with it the advent of capital investmnt in it as an industry and a changed relationship to society in terms of labour demand. Motorization led to an expansion of fishing grounds and the ability to select the grounds to be fished, which caused a dramatic increase in fishery production. At the same time it increased the ability of the fishermen to choose what kinds of fish they would catch.

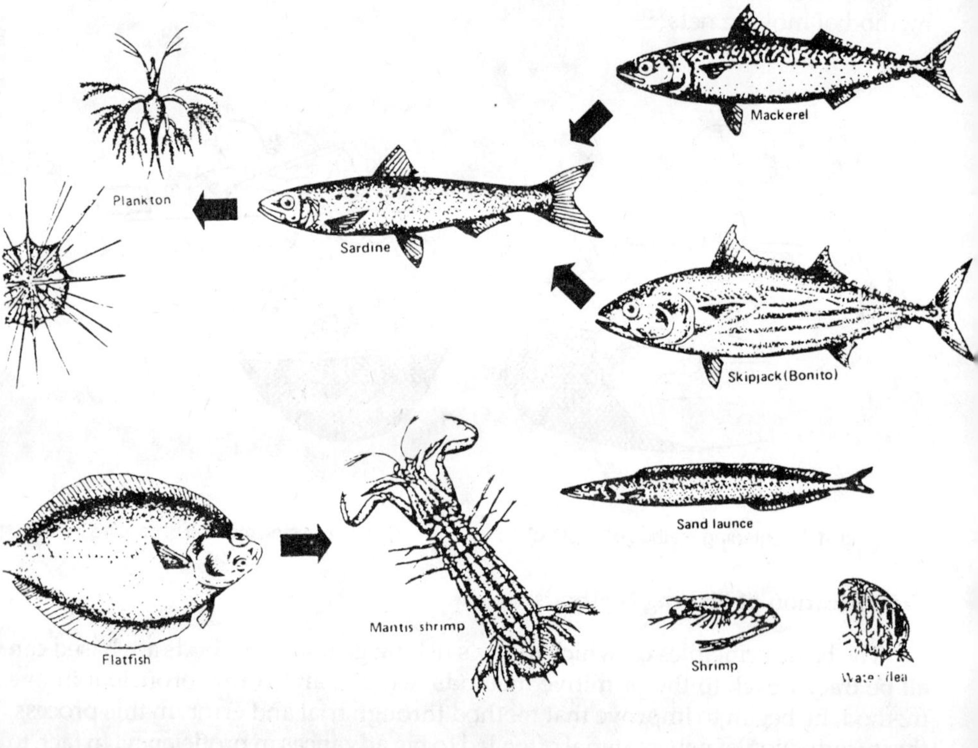

Fig. 6: Horizontal feeding migration

Fish Gathering Devices

For a continuous marketable fish catch, fishermen must either consistently find places where large number of fish is living, or use some methods to gather into one place as large number of fish originally scattered over a large area of water. These two techniques, fish-finding and fish-attracting, the fishermen must develop when undertaking fish harvesting in natural waters. The various fish-finding and fish-attracting techniques that have been developed for each kind of fish are also major factors determining the nature of fishing gear and methods used, and thus can be said to be at the very base of all fishing technology.

Fish Finding

(a) **Sighting—** This is the method in which fishermen find fish by means of visual sightings after arriving at the fishing ground. It is a method that has been used since older times in whaling and marlin fishery. In these fisheries, the whale's blow and the marlins jump become the sign the fishermen look for. Larger fish or marine animals like these migrate over large areas of sea in search of waters of preferable tempeature. The fishermen know the seasonal migrations of these species extremely well and have the experience to judge from the water temperature or from the colour of the tide just where the best fishng grounds will be on a given day.

(b) **Subsidiary signs—** There are other fish-finding methods that rely on recognizing certain other signs for locating large schools of fish. Skipjack, in their migrations over large areas of sea, tend to congregate at certain recognizable points, such as, around drifting timber or drifting sea weed flows or gatherings of large fish or sea animals, like whales, sharks, pilot whales or dolphins. When fishermen sight such signs, they know there is a good possibility that schools of skip jack will be found in the surrounding waters. At times, skip jack will also follow large schools of small fish like sardine, feeding on them heavily. When this happens, the sardine chased to the surface by the persuing skip jack will attract large flocks of sea birds. Experienced fishermen are able to observe the movements of these flocks of birds from a far distance and judge the size and density of the skip jack schools swimming below them.

The shoaling behaviour of sardines during migration exhibit following behaviours, which an experienced fishermen can observe for the presence of sardine shoals in the area:

(i) Surface shoals of oil sardines may exhibit flipping behaviour, when the fish flips out of water being on the surface; pattering of water without coming out of water; making small ripples or waves on water surface or when they leap out of water regularly.

(ii) Bottom shoals, however, do not come to the surface but can make bubbles, which come out to the surface in form of forth; or may secrete mucus having strong and peculiar smell.

The shoals may have bluish or pinkish in colour or may be lumniscent in dark nights.

(c) **Mountain sighting—** Demersal species of fish, shellfish and other marine animals like octopus and sea urchin, either inhabit in rocky bottom areas or prefer sand or mud bottoms. But in either case these animals find the most suitable habitat and repeat migrations between shallow and deep waters in accordance with seasonal changes in water temperature. From long years of experience, the fishermen come to know when the best fishing seasons will be and where the best fishing grounds will be for each speices. In order to confirm the exact locations of their boat at sea, they see lining up a number of landmarks, like mountain points of land or buildings in order to position their boats exactly over the desired fishing ground.

(d) **Scientific methods**— Besides the above traditional methods, which have been developed over long years of experience, modern scientific and technical advancements have led to the development of a new generation of fish-finding instruments that greatly improve the fish-finding effectiveness of fishermen today. Based on the sonar technology developed for military purposes in World War II, these devices (Eco sounder) send pulses of ultrasonic waves down into the water and register the echos reflected back from the sea bottom and other objects in the water on a screen similar to radar. Thus fishermen are able to get not only a picture of the topography of the sea bottom but also identify schools of fish in the deep, middle and surface water layers. From the blips that appear on the eco-sounder screen, fishermen are able to distinguish both the type of fish and size of the school.

With the LORAN position-finding system fishermen tune into a signal broad cast by the regional short wave station and are automatically given a digital read out of their boats position in longitude and latitude.

Recently the appearance of satellite transmission has brought another big change in fish-finding technology. Ultrared cameras on satellites can gather data about the surface water temperatures and then process to create maps of water temperature distribution of the oceans. These maps, which show the courses of warm and cold-water pockets, have become indispensable information sources for off shore, fisheries, such as, large scale seine fishery and tuna fisheries.

Fish Attracting

(a) **Guiding**—Since ancient times fishermen have used techniques to lead fish into their catching devices. In the case of set nets in sea or fish weirs in inland waters, the fence net serves this purpose.

(b) **Startling**—The ground rope on a bottom trawl net or the long wing nets of a boat seine have the effect of startling the fish and driving them toward the centre of the net, either by the sound they make in the water or the flashing of the fibres in the net cord. In one variation of the bottom trawl net, called the otter trawl, the trawl door serves the dual purpose of a spreader to hold the mouth of the net open and also a rake to drag along the sea bottom and stir up clouds of silt that startle the fish schools.

(c) **Light attraction**— The technique of using lights to gather fish was already common in the 8th centuary AD. Fishing methods using fire or other forms of light have existed in all parts of the world since ancient times, and today they continue to be a feature of contemporary commercial fisheries in many countries.

In the Mediterranean countries lights are used to attract pelagic fish like anchovy, mackerel and jacks to be caught by nets. In Egypt's Gulf of Suez there are boat seine and gill net fisheries in which boats with lamps are used to lure sardine and other fish into the nets. In France's Bay of Lyon there is a fishing method in which lamp boats are used to guide fish schools into floating gill nets. In Russia there are seine and scoop net fisheries that use fish-attracting lights, as well as a fishery that uses underwater lights to attract fish to be sucked up with a pump; and another that uses lights to catch fish with a "ship-side pursed net" that combines the qualities of both a purse net and boat seine. The two-boat trawl net fisheries catching herring and cod off the coasts of Sweden and Germany flash spotlights on the water just ahead of the net mouth in order to lure fish into it.

Seine fisheries in Norway use both on-board and underwater lights to attract fish. In America there is a fishery for Pacific squid that uses lights in combination with a suction pump. In Latin America, lights are used to catch anchovy. In Asia, laswag scoop seine, a kind of lift-net fishery of Philippine uses fish lamps. In Korea there is a scoop net fishery that also uses fish-attracting lamps. The lift net fishery of Thailand use fish lights to catch squid. Among all the different kinds of fish-attracting techniques, the use of light is the most widely adopted, contributing greatly to bulk catching of pelagic fish species.

In Japan, the following three fisheries rely heavily on fish-attracting lights:

(a) purse seine fisheries,

(b) stick held dip net fisheries and

(c) squid jigging fisheries.

Although all three of the above fisheries use high-powered lights, the way the light is used is unique with each of the three, creating significant differences in effect.

Phototaxis

The word taxis is referred to the phenomenon in which an animal reacts to an external stimulus with some particular patterns of action. Various forms of taxis can be induced by stimuli ranging from light, water flow and electrical current to chemical agents, gravitational force and any solid objects. In the case of fish, the primary form of taxis observed is phototaxis, or taxis in response to light. With the exception of some species living in the very deep waters, most fish have highly developed optical organs. They are very sensitive to light and dark and also distinguish colours and shapes. Fish demonstrate two types of taxis with regard to light, depending on the species; a positive response, in which they move in the direction of the light and a negative response in which they move away from the light source. Sardine, mackrel and jacks have a strong positive reaction to light, while species like yellow tail, sea bass, catfish and carp have a weak reaction.

Within the same species the response can change with the developmental stage of the life cycle. Eel will gather around light in the fry stage, but show a negative reaction to light when they become adults.

Enough is not known about the phototaxic mechanism in fish. Of the various theories about the mechanism put forth, the two prominent ones are the "Compulsive motion theory" and the "trial error theory".

Compulsive Motion Theory

According to this theory, when fish receive a strong stimulus from a single light source, it causes a disturbance of the balance of light energy they receive from the right and left eyes. In order to recover the balance of their optic organs the fish have a compulsion to either swim directly toward the light source or around it in a circular pattern.

Trial and Error Theory

The theory states that each species of fish has its own optimum light environment. In response to stimuli from the changing environment the fish undertake a variety of exploratory movements in search of ideal environmental conditions. The common point in these conflicting theories is that the fish react to external changes in an attempt to regain their preferred orientation.

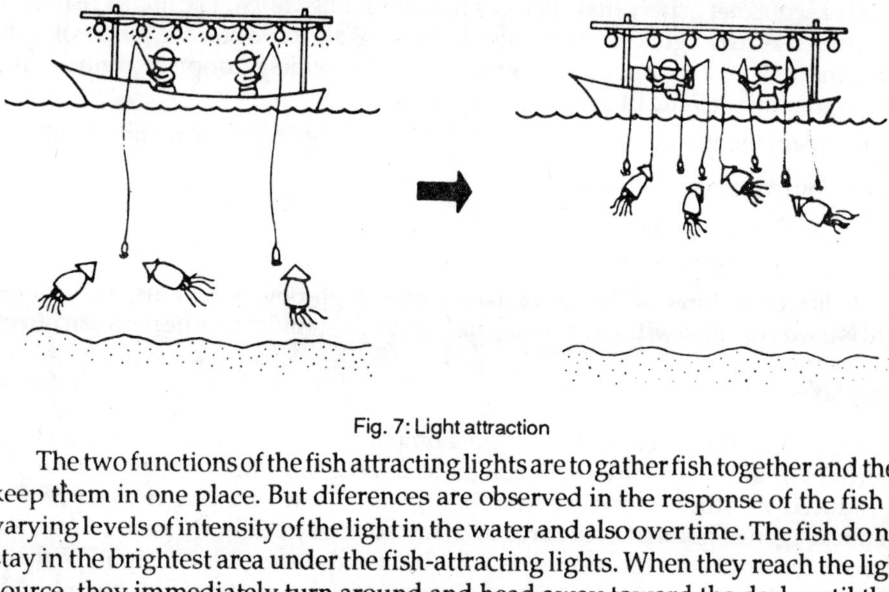

Fig. 7: Light attraction

The two functions of the fish attracting lights are to gather fish together and then keep them in one place. But diferences are observed in the response of the fish to varying levels of intensity of the light in the water and also over time. The fish do not stay in the brightest area under the fish-attracting lights. When they reach the light source, they immediately turn around and head away toward the dark until they turn once again and head toward the light. There is a constant turnover of the fish under the lights, as they repeat cycles of congregation and dispersal.

Many fishermen believe that the brighter the lamps, the higher is the fish-attracting efficiency. However, in tank experiments, using 150 W underwater lamps at 100V, it was found that jack mackerel swam in a circular motion around the lamp at 15–0.1 lux; fish gathered in high concentrations at 2–0.2 lux, the highest concentration being at a level of 0.2 lux.

At one stage in Japan Sea squid fishery, it was discovered that additional increases in the power of lamps did not lead to any increase in catch. Subsequently it was discovered that the largest squid catch was made when the lines were operated in the border area between light and dark areas. Squid have optical organs

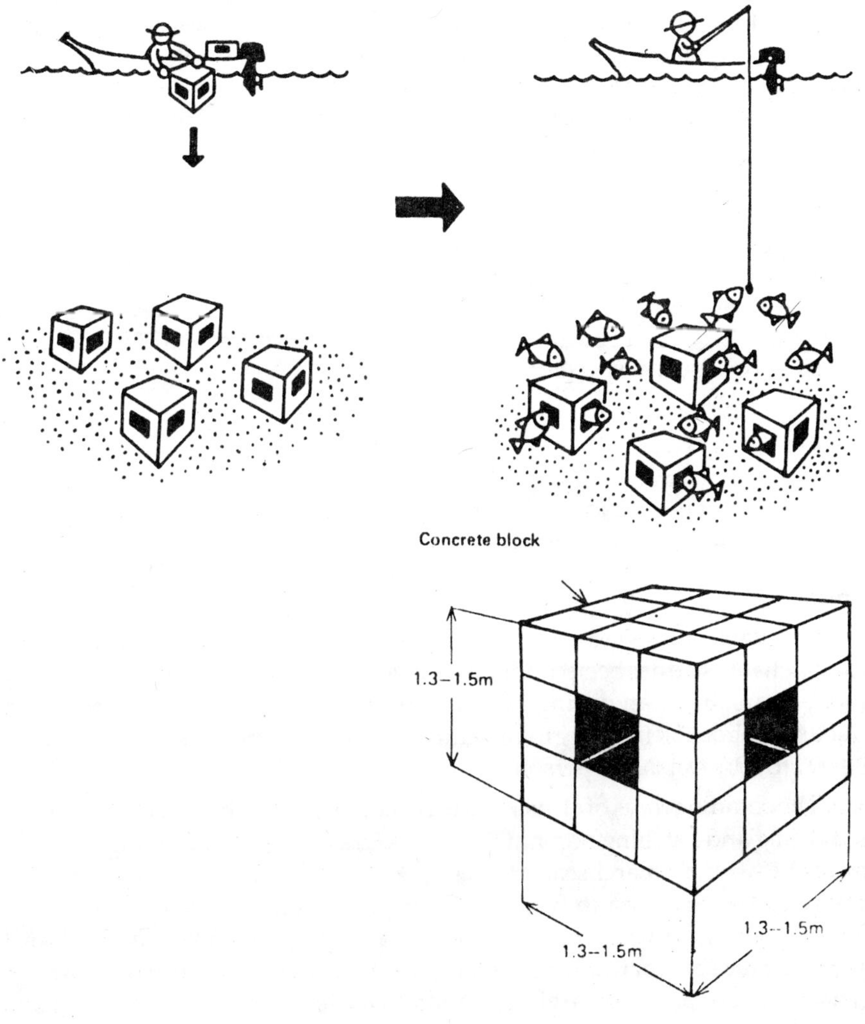

Fig: 8: Artificial reefs

that are highly sensitive to very low levels of light; and the squid's movement is slow and poorly articulated when moving from dark toward light, but is quick and precise when moving from light toward dark. Therefore, the dark area under the boat is very important for catching squid.

Artificial Reefs

Fishermen have long known, and made use of the knowledge, that floating man-made reefs in the water or setting them up on the sea bottom is an effective way to attract fish and other marine animals. These man-made structures that function to gather fish is called artificial fish reefs. In the case of entrapping type fishing gears, like octopus pots, baskets and bamboo tubes, the gear itself functions as a kind of artificial fish reef.

There are number of theories as to why fish gather around artificial fish reef. These include shelter theory, which says that fish choose places where shadows are formed as their hiding places; the feeding theory, which says that the fish feed on the small creatures that attach themselves to the reef, the thigmotaxis theory that says fish like to make physical contact with stationary objects; and the eddy theory, which says that fish are attracted to the eddies that form around artificial reef structures. But it is actually combinations of these various factors that function to produce the fish-attracting effect depending on the species and the specific conditions that exist.

Since 1950, under the coastal fishery facilities development programme in Japan a variety of artificial reef structures of varying sizes and built of a number of durable materials, such as concrete, steel, rubber and plastic have been developed. In recent years there has also been a growth in research and development of floating fish reefs aimed at pelagic fish species.

Bait

In addition to attaching bait to the hook in angling, there is also a technique in which live bait is used to attract fish. In skipjack pole-and-line fishery live sardine or anchovy are chummed into the water to keep a school of skipjack in the vicinity of the boat. Domestication has been practised in many regions of Japan since ancient times.

Regular feedings are used to "domesticate" a population of fish until they grow to a catchable size has been conducted for yellow tail, amber jack and striped jack for many years. In Japan during 1935, this type of "domestication" fishery was conducted in about 50 fishing grounds throughout the region, each yielding a catch of 30,000 to 50,000 fish per person.

The common traits of all fish that become the object of this kind of fishery are that spawning and hatching occur in the open sea, after which the young migrate to the coastal waters of inland seas or bays. Upon reaching the age of one year, the fish return to the open sea to begin cycles of migration. Domestication type fishery involves designating a certain spot at sea where feeds like shreaded anchovy, sardine, blue sprat or mackerel is spread on the water every day to keep a migrating school in one place for a given period, during which time a large number of fishermen make catches from the school by angling.

For the fishing ground, an area of shallow sea with a mild current and a depth of 30 to 50 m. is chosen. The ideal bottom topography is a sunken flat with a white sand bottom surrounded by rocky reef. Such spots are most likely to provide a successful feeding effect, because they are not preferred habitats for demersal fishes and there is reduced chance of reef-dwelling fishes taking a share of the feed.

Fish Aggregating Device (FAD)

Fishing techniques, making use of brush shelters (Payao) have been common in all parts of the world. James Hornell, writes that the origins of brush shelter fishing techniques were observed in India, China, Southwest Asia and Central Africa since prehistoric times. This fishing devices are also referred to in Malaysia as "unjang" and in Indonesia as "rampon".

In this type of fishery, tree branches with full leavage are tied together in bundles and set afloat individually or linked with a longline in places on rivers, lakes or shallow sea waters where fish tend to gather. After a certain amount of time, the bundles are raised slowly and the fish and other marine animals that have taken shelter in among the leaves are shaken out to be caught in a net. In Japan, variations of this method have been used traditionally in certain regions to catch dolphin fish, squid or eel. In 1979, the South Pacific Commission (SPC) conducted successful experiments with payao in the seas around Hawaii. This led to the rapid spread of payao fishery techniques in the countries of the South Pacific.

A payao fishery operation is usually conducted by a single fisherman working from a small 3-5 tonne boat and employing a troll or hand line fishing method. In Japan, the operation of net fishery in the vicinity of payao devices is prohibited. The troll method is particularly effective and has produced good results for Japanese fishermen.

It can be concluded that the payao are effective in attracting the same kind of fish, while eliminating the time and labour otherwise required for searching out productive fishing grounds and, thus greatly increasing the fishing efficiency.

A payao consists of three basic parts, (i) the float structures (raft); (ii) the anchoring lines and (iii) the anchor. In Japan, three private manufacturers, each developed and marketed their own designs. Over time, however, the local fishermen, added their own modifications and improvements to create,what would best be called a next generation of "home made-types" at half the cost of manufacturers.

Numerous researchers have put forth the following theories of explanation for the reasons of payao, as an effective fish gathering device:

(i) Small fish feed on the seaweeds and planktons that grow on and around the payao structure and middle and large size fish gather to feed on the small fish, thus creating an independent food chain centred around the payao.

(ii) The fry and young of the various species gather in the shadows of the payao as a place to hide from predators.

(iii) Fish have an innate habit of gathering in dark areas beneath objects in the sea.

(iv) Some species of fish have the habit of laying eggs on floating masses of seaweeds, etc.

(v) Fish are attracted to the underwater sound given off when weaves break against floating objects or anchoring lines.

According to the studies, the types of fish gather around payao are numerous, with about 30 species being identified. However, the payao's effectiveness in attracting larger fish like skipjack and tuna seems to vary with the location of the payao; with installations set in waters with a depth of over 1000 m. having the highest gathering ratio.

When preparing a payao installation, the most important point to consider is the durability of the device as a whole. Because the device is set in the open sea, there is always considerable danger of the float body being broken up by waves or the anchoring lines being broken by the ocean currents. Generally, a spot on the ocean floor that has steep crevices is chosen for setting the anchor. A payao is usually expected to last for one or two years; but improvements in the materials used in the body structure, anchoring lines and shackles, etc. has increased the durability of some recent models to over two years.

Sound Conditioning

Fish exhibit sensitivity to a certain range of underwater sound frequencies. In recent years efforts have been undertaken to raise fry that are conditioned to respond to a sound signal generated in the water, thus forming the basis for a new kind of conditioned response fish farming. Red sea bream fry are raised in a culturing cage, where they are exposed to the same frequency signal at every feeding in order to condition them to associate the sound with feeding.

After the young are thoroughly trained to gather at the sound of the signal, they are released to the open sea where they grow; but still retain the learned response to gather at the source of an underwater sound transmission of the same frequency.

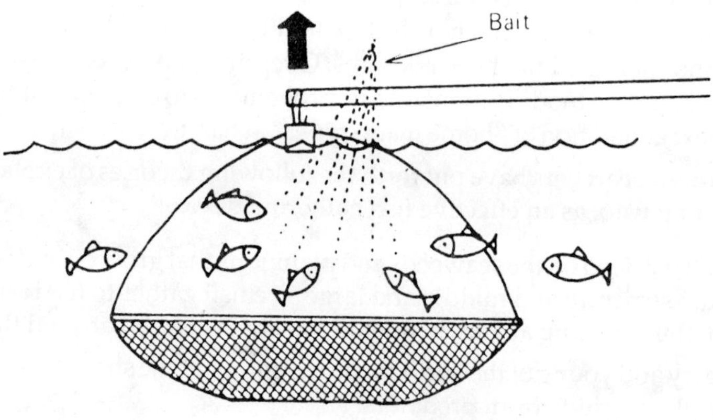

Fig. 9: Bait fishing

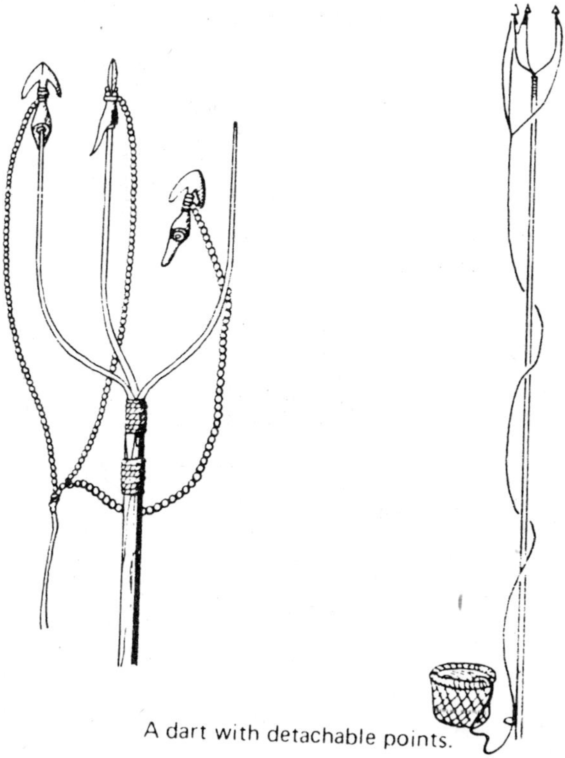

Fig. 10: Spears

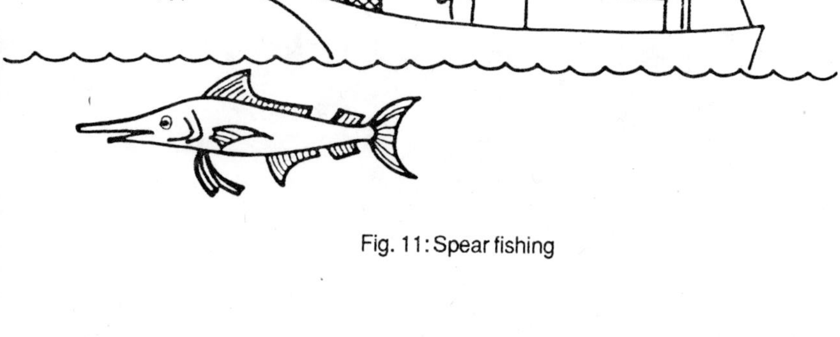

Fig. 11: Spear fishing

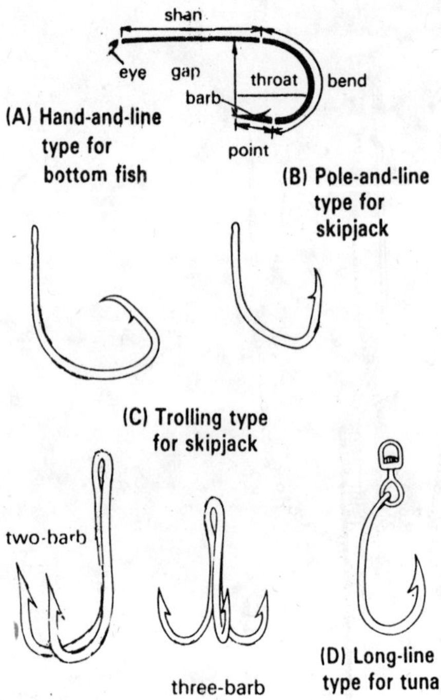

Fig. 12: Fishing hooks

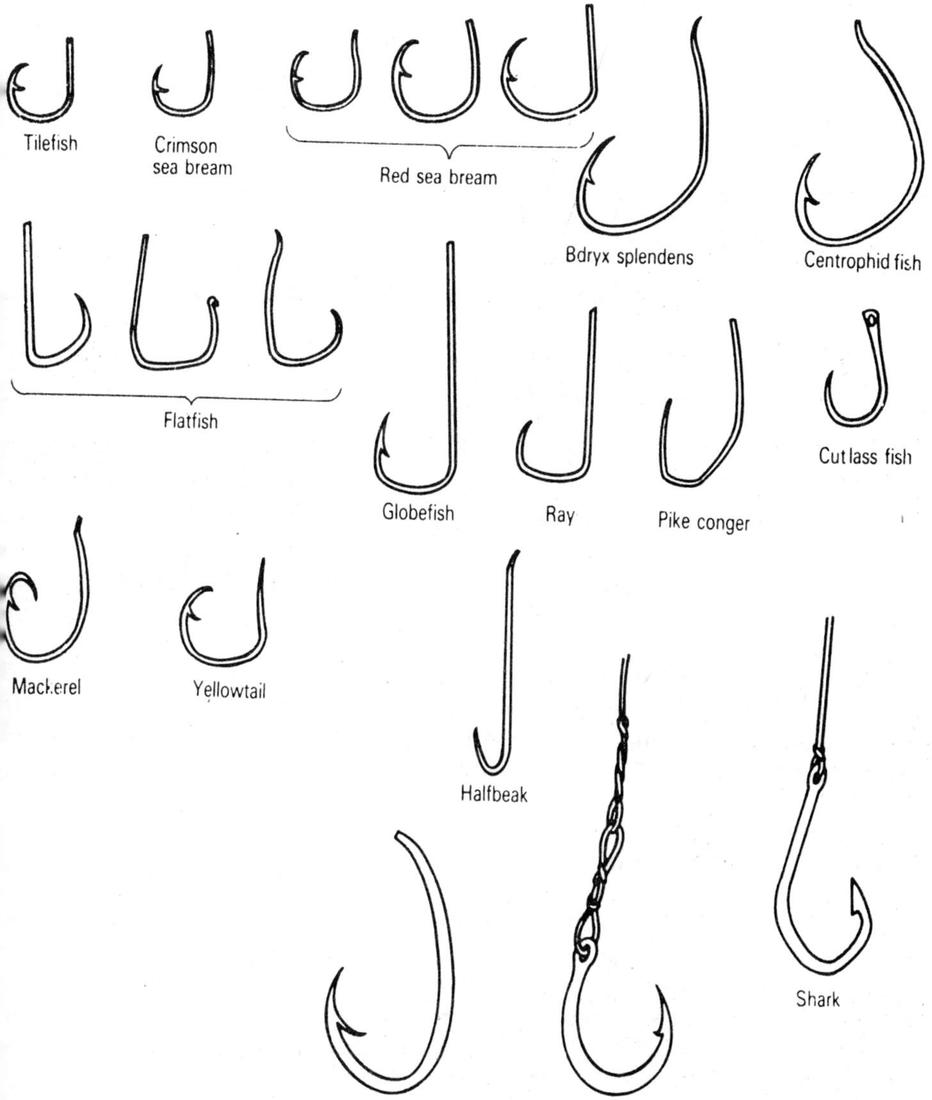

Fig. 13: Fishing hooks shapes

2

TRADITIONAL HARVESTING METHODS

Methods based on Fish Movements

Long ago, when man first began to catch fish, he used the bones, horns or teeth of animals to make two kinds of fishing implements, spearheads and fishing hooks

The spears primitive man–made for fishing were of two types: the fixed type in which the spearhead was fixed firmly to the handle (pole) to be thrust or thrown at fish in the water; and the detaching type in which spearhead would come free from the handle once imbedded in the fish or sea animal. The former was used primarily for spearing fish or shellfish, etc. in the shallow waters of the rivers, lakes or bays The latter was used in the open sea to spear such animals as sea lions, seals and walruses, or larger fish like tuna and billfishes. As time went by, both of these fishing methods gradually went out of use, although they can still be found in certain localities around the world today in specialized fisheries, such as, shellfish gathering, spearfishing and whaling.

Spear Fishing in Tide Pools

At low tide, many fishes left by the ebbing water are swimming in the tide pools of coral reefs. These fishes can be caught by diving into the water with a harpoon or a spear. Shore fishes such as blue humphead parrotfish, thread sail, filefish, and ballonfish are caught, and on rare occasions porgy and striped jack, which come from the open sea also are caught. Octopus and spiny lobster inhabiting the rocky zone can be caught by a special technique using a hook.

Spear Fishing in the Open Sea

Spanish mackerel, inhabiting warm sea areas, has a maximum body length of about 2 metres. They migrate through the open sea and come near the coast in the spawning season (spring). Although Spanish mackerel is easily caught by trolling, they are also caught by spearing, luring the fish with a dummy.

Good fishing grounds are found around bottom reefs. As the fishing boat is set adrift, Spanish mackerel are lured by towing a dummy fish. When the aimed fish comes near the dummy, the dummy is pulled in a way that makes it jump up into the air at the right time. The fisherman then throws a spear to catch the fish just as it starts to jump up after the dummy. Since a rope is tied to the spear, used in this method, the shooter catches the fish by hauling the rope after the fish has become inactive.

Fishing Hooks—The Oldest and Most Highly Perfected Fishing Gear

The use of fishing hook, unlike spear, on the other hand, has continued to grow in increasingly effective forms as man has accumulated experience and knowledge about the behaviour and life patterns of fishes. Especially since man entered the Iron Age and became proficient at making fish hooks of metal, he has continued to improve and rationalize hooks into an increasing variety of shapes for each type of fish to be caught. Particularly in the beginning of the Iron Age, considerable efforts were devoted to perfecting the shapes of fish hooks in ways that increased their performance dramatically.

There are several different ways a fish may take a prey or bait, such as, an attack action, where the fish approaches its prey rapidly and bites into it; or sucking action, where it approaches slowly and cautiously and then sucks it in. But in either case, there is a common process to the act of taking in a prey or bait. The mouth is opened as the fish approaches the prey and then sucks it in along with the water around it with the gillcovers open wide on both sides. Once inside the mouth, the water is forced out through the gills and the prey or bait is taken into the stomach cavity.

The fishing hook is an implement that takes advantage of this sucking action to catch the fish. First the hook with bait attached to the bend portion hangs in a balanced position in the water. Then, when the fish sucks in the bait and hook, the gravity of the shank causes a shift in that balance that sends the point into a circular motion, with the eye as the pivot point, causing it to stick into the upper jaw of the fish.

The shape of a fishing hook varies with the type of fish it is intended to catch, but the basic shapes can be roughly divided into three types: angular, round and elongated types. These groupings seem to be more an expression of the regional traditions that grew out of the way the hooks were used, rather than any specific technological differentiation.

The reason that the fising hook with its essentially passive nature was able to win universal acceptance over the essentially active spear as a fishing gear, is because of the following advantages that angling had as a fishing method:

(i) Fresh or live bait or artificial lures could be attached to it to attract the fish.

(ii) By attaching a sufficient line to it, fish not only in the surface waters but also the middle and deep water layers could be caught.

(iii) With the addition of accessories like swivels, multiple–hook fishing gear could also be prepared.

(iv) By using the long line method, large areas of water could be fished at one time.

(v) Wind power, current flow or mechanical power could be used to conduct trolling fishery.

(vi) By varying the type of bait used and the size of hook, selectivity could be achieved with regard to the type of fish to be caught.

Line Fishing Methods

Line fishing is mostly operated on a small scale in coastal waters; the exceptions are tuna longline and skipjack pole–and–line fishing. The construction of line fishing gear and the methods of operation vary according to the fishing ground, fishing season and fishermen's preferences, even for the same kind of fishing.

Line Fishing Gear

In comparison with the gear for net fishing, the line fishing gear is simple, being composed basically of a line and a hook. Other elements, such as pole, float, sinker, swivel and snap are added to the gear required. The amount of catch is very much influenced by the efficiency of gear; therefore, the choice of material and the construction of gear should be done carefully.

(i) Hook — Hooks for line fishing exist in a variety of shapes and sizes. The materials of which hooks are made are mainly iron (galvanized or tinned for protection against rust), brass and stainless steel.

According to their shapes, hooks can be classified into three types; round, angular and long type. Besides, there are complex hooks, such as, double and triple hooks, and hooks elaborately constructed as jigs for squid and for tuna pole-and-line fishing.

A hook consists of five distinct parts, namely, head, shank, bend, point and barb.

The function of the hook head is to allow the line to be fastened easily and securely to the hook. A hook with a long shank can be handled easily; it is easy to place the bait on the hook and to remove the fish from the hook. A long shank also, prevents the line from being cut by fishe's teeth. But the catching efficiency of such hooks is not high.

In order to prevent the captured fish from dropping out of the hook, most hooks have a barb. In some cases, when it is necessary to remove the fish from the hook quickly, as in the case of pole-and-line skipjack fishing, hooks without barbs are used. The hook with the point bent inwards is the most suitable for use with bottom longline as it does not enter the fish's body very deeply and thus allow the fish to remain alive for some time after swallowing the hook.

The hook twisted in the bend part is convenient because the bait can be placed on the hook easily and its catching efficiency is high.

Double and tripple hooks are used for troll fishing; these hooks are effective to prevent the fish from dropping out of the hook.

Squid jigs vary in shape, colour and materials. Body parts are made of plastics, vinyl, rubber, cloth, wood, lead etc.

Bonito and tuna jigs also vary in colour and materials of the lure part (body part). Generally fish skin and chicken feathers are used to conceal the hook. It was found that the smaller the hook, the higher its catching efficiency.

(ii) Line — The material and thickness of the line depend on what kind of fishing it is used for. In case of tuna long-line, the main line is 6 mm. in diameter and is made of vinylon, nylon, polyester etc. Branch lines are somewhat thinner and are also made of nylon, vinylon, or polyester, whereas the hook line is made of steel wire. In case of handline and pole-and-line fishing, the main line is not as thick as that of tuna longline. Nylon or polyethylene are used for both the main line and the hook line.

Generally, lines for the line fishing should be as thin and as strong as possible, particularly branch lines and hook lines. So the nylon monofilament is the best material for line fishing.

(iii) Pole — Commonly, angling pole is used to catch the fish swimming near the surface. There are several advantages in using a pole:

(a) The hook can be moved to the desired position easily.

(b) The shock caused by the fish can be mitigated, thus preventing the line from breaking.

(c) The fish captured is taken on board easily and quickly.

Bamboo used to be the most popular material for the pole, but recently, fibreglass pole has been developed and is now widely used.

On the other hand, in order to catch the fish swimming in deep waters by hand line, a special device, "Balance" is used. A piece of thin stainless steel wire with a strong spring attached to the end of the line has the function of a pole.

Strength of nylon monofilament line

Diameter in mm.	*Breaking strength in kg.*	*Knot strength in kg.*
0.29	4.2	2.7
0.37	7.3	4.1
0.44	10.7	5.9
0.52	14.4	8.0
0.57	17.0	8.1
0.66	21.9	10.4
0.70	24.0	10.9
0.74	27.1	12.2
0.81	30.7	13.5
0.87	36.4	16.6
0.90	38.1	17.8
0.97	41.4	19.8
1.04	46.9	21.5
1.17	56.9	27.6
1.28	65.1	28.4
1.38	74.5	30.0
1.47	82.0	35.0
1.65	101.0	47.0
1.81	115.0	57.2

(iv) Floats — The float serves the following purposes:

(a) To place the hook at the desired position (depth).

(b) To indicate whether the fish is hooked.

(c) To mark the location of the gear.

In commercial fisheries, float is used mainly for long line fishing. A barrel, a block of wood, an empty pot etc. are still commonly used in small-scale coastal fishing, but glass or polyethylene floats are used in large-scale fishing, such as, tuna longline in deep waters.

(v) Sinkers — Sinker has the following roles:

(a) To sink the hook to the required depth quickly.

(b) To prevent the hook and line from drifting away by current.

Stone, iron, lead etc. are used as sinkers.

(vi) Swivel — Swivel is used in joints, e.g. between main line and branch line, or the branch line and the hook line, to prevent the line from twisting.

There are many types of swivels, but the type and the size are decided according to the kinds of fishing and size of fishing gear.

(vii) Snap — Snap has been developed for tuna longline fishing to join the branch line to the main line easily. Now-a-days, the snap is used widely not only for tuna longline, but also for some other kinds of fishing, such as, trap fishing.

Classification of Line Fishing Methods

Line fishing is classified according to the type of gear construction as follows:

- Line fishing
  - Handline
  - Pole-and-line
  - Longline
    - Drift longline
    - Bottom longline
  - Trolling

Handline Fishing

Handline gear is very simple. It is composed of the main line, branch lines, hook and lead. Some types of handline gear have several hooks, e.g. a main line with an end-lead may have hooks on branch lines so that vertical long lines are formed. Fishing with a "Balance" is also considered a variation of hand line fishing. Handline gear is used to catch the fish dwelling in midwater and near/on the sea bottom. Sea bream, mackerel, jack-mackerel and other bottom fish are caught by handline fishing.

Handline Fishing for Demersal Fish

Saury fish, squid etc. are used as bait. One filet of saury fish is placed on the hook through the tail part. Squid is cut into strips of about 30 cm. in length for use as bait.

In this method, a 3-5 tonne boat with engine power of 15-45 h. p. is used with 2-4 crew on board. A fish finder is used to locate the school of fish. On finding the fish school, the boat should sail above the fish school, against the current, and the gear is shot. After the boat has drifted across the fish school, it should be brought to its initial position above the school of fish before the fishing operation is repeated.

Handline Fishing for Mackerel and Jack-Mackerel

Cat skin, trigger-fish skin, thin strips of rubber etc. are used as bait.

A one to three tonne boat is used for operation, with one fisherman on board. The boat is equipped with fish luring lamps, powered by a 24 V, 1 kW generator. The boat is anchored above the fish bank after sunset, two fishing lamps of 100-300 W are hung above the water, one on the either side of the boat. The fisherman operates handline gear and pole-and-line gear. The operation is carried out until sunrise. This type of fishing is suspended for three days during the full moon period.

Balance Handline Fishing for Mackerel and Jack-Mackerel

The balance functions as a pole under water. It is usually made in the shape of a crescent. With the up-and-down movement of the fishing line, the balance vibrates causing vibration of the hooks.

Handline Fishing for Squid

Fishing operations are conducted on board of a 0.5–3 tonne boat, by 1–3 crew. The boat is equipped with a fish luring lamp which hangs in the centre of the boat, powered by a battery, or sometimes by a generator as in the case of larger boats. There are two fishing methods: one is for the night-time fishing and the other is for day-time fishing. The squid fishing is generally done in the night.

On arriving at the fishing ground, the boat is anchored. The hooks are at first shot near the sea bottom, but gradually, as the squid are lured to the surface, the fishing line is shortened and the hooks brought close to the surface.

With the modernisation of fishing boats (increased size of boats, mechanization, etc.), the gear has also improved. A notable recent development been the introduction of the automatic squid angling machine.

Of various coloured squid jigs, the red and orange jigs gave good catches than the green and luminous jigs. The fishing results also indicate that the plastic triple-decker jig has the highest catching efficiency. But this type of jig is easily damaged with its long hooks.

Using fish luring lamp, the best catch was obtained when the gear entered the sea at the borderline between the areas of light and darkness. The angling gear placed in the shadow zone obtained better catch than the gear which was submerged in the illuminated area.

Pole-and-line Fishing

Pole-and-line fishing gear is used mostly to catch the fish swimming near the water surface.

The advantages of pole-and-line fishing, in comparison with other line fishing methods are: taking the captured fish into the boat is done easily and quickly and the shock caused by hooked fish is lessened.

Pole-and-line fishing on a large scale is carried out mainly for skipjack and mackerel.

Skipjack Pole-and-line Fishing

Skipjack is an important species in commercial fisheries in tropical areas. The species is distributed widely in the warm waters of the world and is caught mainly by purse seine, pole-and-line and gill net.

Siting of skipjack school can be done by means of a fish finder, trolling with lure, or by visual observation. Skipjacks form schools around drifting objects, whale sharks and baleen whales. Flocks of birds often circle above schools of skip jack and feed on small fish.

When the school of skipjack is spotted near large drifting objects, whales, sharks or a circling flock of birds, the boat should approach the school and live bait fish is thrown from the bow to attract the school toward the boat. Water is also sprayed around the boat, so as to conceal the shadows of fishermen from the fish. Live bait fish is indispensable for skipjack pole-and-line fishing.

Fishing for skipjack often means catching tuna and albacore as well as skipjack For this reason, each boat usually carries 5-8 types of poles, varying in size as well as in construction. For different sizes of fish and schools of different characteristics, fishermen use the appropriate type of pole.

Boats of 30-50 tonnes have 20-25 crew members on board. At the fishing ground, the crew first have to find out the school by means of a trolling line or by watching for drifting objects, whales, sharks, jumping skip jack and birds. When the school of skipjack is sighted, the boat approaches it, live bait fish (anchovy, sardine, pilchard, herring) is thrown from the bow, while water is sprinkled from the sides of the boat. If the school of skipjack is attracted around the boat, boat is stopped and the operation begins. Generally, operation is done at the port side of the boat, and this side should be kept on the lee side.

Normally unbarbed jigs are used for skipjack pole-and-line fishing, when the captured fish drops off the hook automatically when the line is slackened and the operation is done more quickly than with a live bait hook. 1.7 second on average with maximum 3 seconds are required to catch a fish by this method. Jigs with white chicken feathers are good for catching skipjack and the jigs with white and brown chicken feathers are used for albacore pole-and-linc fishing.

On the other hand, when the boat finds a school of skipjack which can not be caught by jigs, live bait hooks are used. Sometimes, skipjack is caught by jigs as long as the catch is good, but when the catching rate decreases, live bait hooks are used instead of jigs.

Anchovy, 6–10 gm. (6-10 cm.) in size, is suitable as the bait fish for skipjack pole-and-line. Larger anchovy, 8-15 gm. (8-13 cm.) in size is used for albacore pole-and-line fishing. A 40 tonne boat uses 80-120 buckets (18 litres capacity) of live bait fish in the course of one fishing trip.

Longline Fishing

Handline and pole-and-line fishing is active fishing operations, done at the place where the school of fish is found. On the other hand, longline fishing is passive, that is, the gear is set at the place where the school of fish is expected to be found. So the scale of gear is comparatively larger than the gear of handline and pole-and-line fishing. Two types of longline fishing is done; one drift longline and the other bottom longline.

Tuna longline is the largest in size among drifting longline. Tuna longline fishing is operated in any sea water between the latitudes of 40° south and 40° north. The number of units of gear used is different according to the size of the boat, but the construction of gear is almost the same in all cases.

Tuna longline fishing is operated in the coastal and off shore waters on a small scale with small boats of under 20 tonnes. Generally, tuna caught by small boats and preserved by means of ice and sea water is more expensvie than tuna caught and frozen for long periods on board large vessels which operate in pelagic waters.

Specifications of tuna longline fishing gear

Name	*For Bigeye and albacore*			*For Yellow fin and marlin*		
	Material	*Size/Quantity*		*Material*	*Size/Quantity*	
Main line	Vinylon	5.5 mm.	37.5 m. 14 Pcs.	Vinylon	5.5 mm.	52.6 m. 6 Pcs.
Branch line	"	"	10.5m. 5 Pcs	Polyethylene	3.5 mm.	12 m. 5 Pcs.
Sekiyama	Wire with cotton	3 m.		Nylon		4.5 m. 5 Pcs.
Hook line	Wire	2 m.			Wire	3m. 5 Pcs.
Swivel						
Hook						
Float line	Polyethylene	5.5 mm.	12 m.	Polyethylene	3.5 mm.	10.5 m.
Float	Glass	24 cm.		Glass	18 cm.	
Radio buoy			3–4 Pcs.			3 Pcs.
Light buoy	Battery	12 V.	6-7 Pcs for every 15-20 baskets	Battery	12 V.	6-7 Pcs. for every 20 baskets

One unit (one basket) of the main line for catching bigeye and albacore is composed of 14 lines, each one 37.5 m. long. Branch lines are joined to the joining parts of the main line.

In case of yellowfin tuna longline and marlin longline, the main line consists of 6 parts, each one 52.6 m. long. The yellowfin tuna long line has 5 branch lines, whereas the marlin longline has 4 branch lines and the float line is attached to the centre part of the main line.

A branch line is composed of 3 parts—the branch line proper, sekiyama and the hook line. A swivel is used between the branch line and sekiyama to prevent the hook line from twisting and entangling with the main line. A hook is fixed to the hook line by means of snap.

Fishing boats of less than 10 tonnes, 70-100 h. p, with 6-7 crew are used. Each boat has a live bait preservation hold, a fish hold with ice, line hauler, side roller, fish finder, wireless, and Loran.

Live jack-mackerel is used for yellowfin and marlin longline fishing mainly, but frozen saury, squid and mackerel are also commonly used. In the case of live bait fish, the hook pierces slightly the front part of the first dorsal fin, but when frozen bait fish is used, the hook pierces the head of the fish from top downwards. The choice of fishing ground is made before departure of the boat, according to the data obtained from the fishing information centre or other boats in the port; but the exact operation position is decided by sea conditions, such as, water temperature, water colour etc., and the presence of other boats operating in the area.

Lines are shot around sunrise from the stern part of the boat, while the bait fish are placed on the hooks. It takes about 3 hours to shoot 170 baskets.

Fishing crews during fishing operation should perform the following duties:

(i) shoots the main line and hands over the branch lines;

(ii) to place the bait fish on the hooks and shoots the branch lines;

(iii) to join the float line to the main line and shoots;

(iv) carry the baskets;

(v) prepare the bait fish; and

(vi) steer the boat.

After shooting the line, the boat drifts about 4-5 hours near the end of the line. While the boat is drifting, the crew take a rest and prepare fore hauling. Hauling starts in the afternoon, with the help of line hauler. It takes about 7-8 hours to haul all lines. During hauling the crew do the following works:

(i) handles the line hauler and processes the catch;

(ii) hauls the branch lines, coil them and places them with the main line;

(iii) manages the main line;

(iv) makes the baskets and carries them to the stern part for next shooting;

(v) manages the floats; and

(vi) steer the boat.

Large-scale Tuna Longline Fishing

Tuna longline boats of 200-300 tonnes operate at the pelagic tuna fishing grounds all over the world. These boats are undergoing rapid modernization of facilities and mechanization of fishing equipment in the face of increasing demand for fish as well as rising cost of labour. With the introduction of power equipment, such as, the line winder system and the reel system; it has been possible to reduce the number of crew of a tuna longliner from 27 or 28 to between 17 and 21. There has been a considerable development in the processing and the catch-preservation facilities on board tuna longliners. In modern vessels, the temperature of the freezing chamber is –55° to –60°C, and that of the fish hold is –45° to –50°C, thus allowing excellent preservation of catch for very long periods. At present, tuna longliners operate for 8–12 months continuously, storing their catch on board, only stopping at a port once every 2-3 months to renew their supplies of water and provisions.

The main line can be a single line up to 1,50,000 m. long. When not in operation, it is either wound on the reel on the stern part of the boat which is equipped with the reel system, or it is coiled in the storage box in case of boats equipped with the line winder system. Branch lines are attached to the main line, at suitable intervals, by means of snaps.

All tuna longline boats over 100 GT are built with a long poop deck, which has the advantages of protecting the boat from the following wave or beam wave during shooting of the line. Also, when the line is hauled against the wind at an angle, it is easy to keep this angle, because the resistance of the wind in bow part of the boat is less than in the stern part.

Frozen saury, squid and mackerel are used as bait. The main line of the old type, which consists of many parts, is shot while the line is being joined by baskets, at a boat speed of 5 knots or more. The modern main line is a single piece of line to which branch lines have to be attached, while shooting is done, at the speed of boat about 10 knots.

Shooting generally begins before sunrise. The boat runs with a constant course following the direction of the wind. Hauling begins in the early afternoon with the help of line hauler. Ideally, the course of the boat should be kept at the starboard bow wind. It takes about 10 hours to haul the entire line.

Bottom Longline Fishing

Bottom longline is used widely throughout the world. There are many types of bottom longline, but many of them are operated only on a small scale, generally in coastal waters.

Cotton and vinylon are widely used for the main line, because these materials have high specific gravity, which gives them high sinking speed and reduces the effect of current upon gear. Thickness of the main line is determined by many factors, such as, character of the bottom, speed of the current, depth of water, size of fish, pulling power of fish, number of branch lines and size of the fishing boats. In small boats (1-3 tonnes), operating in waters of less than 100 m. depth. 1.5 mm. dia main line is used; while for boats of 3-10 tonnes, diameter of the main line ranges from 1.5 to 2.2 mm. When fishing is done in waters with strong current, main line thicker than 2.3 mm. diameter is used, regardless of the size of the boat. In most cases, length of one unit (basket) of main line ranges from 200 to 700 m.

Nylon with high flexibility, good breaking strength, elongation and durability is the most suitable material for branch lines, since they should have low visibility and should be moved by current. Thickness and length of branch lines are determined by the size and power of fish and the speed of current. Length will have to be chosen so as to allow easy handling of the gear. Length of branch lines and intervals between branch lines are determined by the size of fish and density of school. Generally, the length of interval between the branch lines is from double to triple the length of a branch line.

Three main types of bottom longline fishing gear is distinguished; the most commonly used type, in which the main line is laid on the sea bottom, having the disadvantage of main line or branch lines getting caught by rocks, resulting in damage or loss of gear. Also bait is taken away from the hooks by bottom dwellers, such as, starfish or crab.

The main line, branch lines and hooks do not touch the sea bottom in second type of bottom line gear, the structure of which is complicated than the other type mentioned above. The risk of the gear caught by rocks or other obstacles at the bottom is not so great and fish which dwell near the bottom are caught by this gear.

The third type is the most complex, in which the main line is suspended at some distance above the bottom, depending on the length of branch lines. Thus the main line does not touch the bottom, whereas the branch lines do. If a branch line gets caught by some obstacle at the bottom, only that particular line would get damaged or lost, without a risk to the rest of the gear. In this type of gear, hooks are placed between the sea bottom and the main line so that the fish which dwell near or at the bottom are caught.

Squid cut into rings or strips is used as bait. Bait is placed on hooks and the gear is arranged in baskets before arrival at the fishing ground. Sinkers are hung on the hooks at 13-15 branch line intervals. 8-10 baskets are shot for one operation. If there are many puffer fish at the fishing ground, buoy should be attached to the main line at 2-3 basket intervals to prevent the main line from being lost, because the puffer fish may cut the main line.

When shooting of the gear has been completed, the boat should drift near the end of the main line. Hauling begin after 15-20 minutes. If the catch is good, a buoy can be used to mark the location so that operation can be carried out again in the same place. 30-40 baskets are used in a day, and so at least 50-60 baskets should be prepared on board.

A 3-4 tonne boat is used for the operation, with the crew of one or two fishermen. In the case of one-man boat, the line hauler is set at the stern, but if there are two fishermen on board, it is set on the bow part of the boat. The line hauler is driven by the main engine through the counter-shaft.

In case of sea-bream bottom longline, squid cut into rounds is used as bait. Shooting is done from the stern part of the boat while bait is put on hooks. The boat should be operated across the current in order to prevent branch lines entangling with the main line.

Hauling begins 30-60 minutes after shooting, by means of a line hauler, which is driven by the main engine. Operation is carried out in the daytime, from sunrise to sunset. 10-15 baskets are used for one operation and 2-3 operations are done in a day.

Saury fish, cut into strips, is used as bait in rockfish bottom line operation. Two-thirds of all branch lines in one basket are laid on the sea bottom and the rest of the branch lines are hung above the sea bottom. Shooting is done while placing baits on hooks. After shooting the boat returns to the position from which it started and hauling begins.

Salted sardine is used as bait in sea bass bottom longline operation, which starts in the early morning. Ten baskets are prepared and shot at ten places separately in order to cover a wide area of the fishing ground.

Squid is used as bait in rockfish vertical bottom longline in the night-time operation and mackerel, saury, sardine, etc. are used in day-time operation. Shooting is done just before sunrise, while hauling begins from the starting point of the line as soon as shooting is finished.

The hooking rate in bottom longline (number of fish caught/number of hooks) is 0.26 in evening operation while, 0.24, 0.13 and 0.09 are in morning, day-time and night-time operations respectively. Hooking rate tends to increase with the longer immersion time, but it is observed that there is a variation in the catch even with nearly the same immersion time. The hooking rate of useful fish (rock cod, rock trout, sea bream and grouper) tends to decrease with the longer immersion time, and the catch of unuseful fish, such as, shark, ray, puffer-fish increases. Also useful fish caught early drop from the hooks or are eaten away by predators. It was observed that grouper were caught by hooks on the upper part of branch lines in morning operation, but in evening fishing they were caught by lower hooks. Rock trout were always caught by lower hooks. Rock cod were caught by lower hooks in morning, evening and night-time operations, but in the day-time they were caught by both lower and upper hooks.

Troll Line Fishing

Troll fishing gear is composed of a line, a trolling jig, and a diving board or splashing float whose function is to submerge or to move the jigs. The trolling jig moves like a small fish so that large fish are lured and captured. Boats used in troll fishing are generally small in size. Each boat tows serveral lines at a speed of 3 to 8 knots. The main species of fish captured by trolling are the pelagic ones, such as bonito, tuna, yellow-tail etc.

There are many shapes of diving boards, but they all have basically the following functions:

(i) to submerge the trolling jigs to the aimed depth;

(ii) to move the jigs so as to stimulate live bait fish;

(iii) to indicate whether the fish are hooked; when this happens, the board comes up to the water surface.

The purpose of the splashing floats, which are of many shapes are:

(i) to move the trolling jigs as if these jigs are real bait fish;

(ii) to create waves and noise similar to those made by small fish;

(iii) to indicate whether fish is hooked; and when fish are hooked, the float will submerge.

The catch is influenced by the type of trolling jig used. Squid-shaped or octopus-shaped lures are used for skipjack troll fishing. The lures are made of vinyl, chicken feathers, fish skin etc. The head of the lure is made of lead or lead inlaid with shell, cow horn, animal bone, ivory or plastic. The hook is fastened to the line.

One to eight tonne boat is used, but the number of crew are one or two only. When the first fish is hooked, it is not taken in, but left on the hook to lure other fish. If the fish are hooked by the lines on only one side of the boat, the boat will change its course to that direction.

In Spanish mackerel trolling gear, 25 pieces of lead are attached along the main line 2 at intervals which become shorter the nearer they are to the hook. The weight of each piece of lead is 45 gm. Four pieces of some shiny, light-reflecting material, such as, abalone shell, are inlaid in the head of the lure. This is believed to increase effectiveness of the lure. The teeth of Spanish mackerel are sharp and can easily damage the line. The hook line should be carefully examined every time when Spanish mackerel is caught. Engine revolution is kept to 70 per cent of normal operation while fishing.

Drive in Net Fishing

This fishing method is operated jointly by group of fishermen from one village. In this method, a net, consisting of a long-wing net and a bag net is spread out beforehand in a certain place. The fishermen then enter the water and scare the fish out from the rocks and drive them into the net with their own hands, feet and a scare line. Some of the fish get entangled in the wing net and a considerable numbers of fish are caught in the bag net. The structure of the netting gear and fishing method vary in accordance with the species of fish being caught.

Barrier Fishing

Fishes and other marine animals are caught by making use of the ebb and flow of the tide. Spreading of nets are made in various ways, e.g. straight, zig–zag, curved and with bag net. At high tide, the net is set in a tide land or a lagoon, which is covered by sea water during high tide. Fishes which come into the net at high tide are entangled in its meshes or trapped within the bent sections of the net during ebb tide.

3

MODERN HARVESTING METHODS

Technical Distinctions in Modern Angling Fisheries

The technology of modern commercial angling aimed at catching the largest possible number of salable fish effectively and economically.

Although there are number of factors involved in angling technology, there is a basic system by which a particular technology is evolved (Table 1) that includes both the knowledge gathered by researchers and the acquired skills of the fishermen combined with knowledge gained from experience.

Commercial angling fisheries are divided into three major and eight sub-categories (Table 2). The major categories are divided according to types of fishing gears, while the sub-categories are divided according to the life habits of the fish to be caught.

Individual skill of the fisherman is the biggest single factor in the success of angling, being directly related to the fishing results. But in case of long-line and automatic angling, this is not necessarily true, as individual skill plays only an indirect role in the final fishing results. In either case, both of these methods are the result of a "standardization" of the fish-catching mechanism. Angling can be further classified into four sub-categories, namely, hand-and-line, pole-and-line, vertical line and trolling.

Hand-and-Line

Fishing line is worked by hand to bring up the catch, is used mainly for catching fish which inhabit the sea floor or fish which migrate in the sea bottom. The most important point in successful angling is to accurately determine the depth at which the fish is migrating. It is also important to know the sea bottom topography of the fishing ground as most bottom-inhabiting fish prefer certain type of sea environment such as, around the edges of the reefs, or sand and mud bottom areas.

In angling there are two basic fishing techniques, one in which the line is moved gently through the water to attract the fish to the baited hook, and another in which the line is hung perpendicularly in the water and remains still in one position. Both the techniques may be used alternately for catching the same type of fish, depending on the season. In case of sea bream, the former method is used in spring and autumn, when the fish are migrating constantly between the deep and shallow waters; while the later is used in summer and winter, when the sea bream remain in bottom waters.

Table 1: System for angling technology

Object	*Condition*	*Connotation*
CHOICE OF FISH TO BE CAUGHT	Resources	• Life habits (Life cycles and feed resources) • Biological environment (Water temp., water quality and sea bottom quality)
	Sea Conditions	• Current flow, Wave conditions, climate
↓	Capabilities of fishing boats available	• Running capability (Wave tolerance, accommodation speed) • Fish finding capabilities (Loran, fish finder, wireless radio) • Operating capabilities (Mechanization of fishing operation and fish hold)
TYPE OF FISHING GROUND	Habitat of schools	• Tendency to inhabit one location • Migrating tendencies (Spawning and feeding migration)
↓	Density and movements of fish schools	• Schooling movements, daily movements (or feeding) • Travelling characteristics (taxis)
CHOICE OF FISHIING GEAR AND FISHING METHOD	Fish attracting activities	• Spreading bait in water. Fish lights. Showering water • Underwater sound. Use of artifical reefs. FAD
	Fishing gear configuration and setting	• Depth of setting. Single hook or multiple hooks (vertical or horizontal line setting)
↓	Movement of fishing gear	• Still or set in one position • Drifting • Towing
FISHING OPERATION	Signal stimulus	• Jigging and drifting with current. Bait, lure
↓	Instinctive of reflective	• Nibbling at the bait • Attacking
CATCHING		

Table 2: Classification of angling fisheries

Major categories	*Sub-categories*
Angling	Hand-and-line
	Pole-and-line
	Vertical line
	Trolling
Long-line	Drift long-line
	Bottom long-line
Automatic angling	Automatic squid jigging
	Automatic skipjack jigging

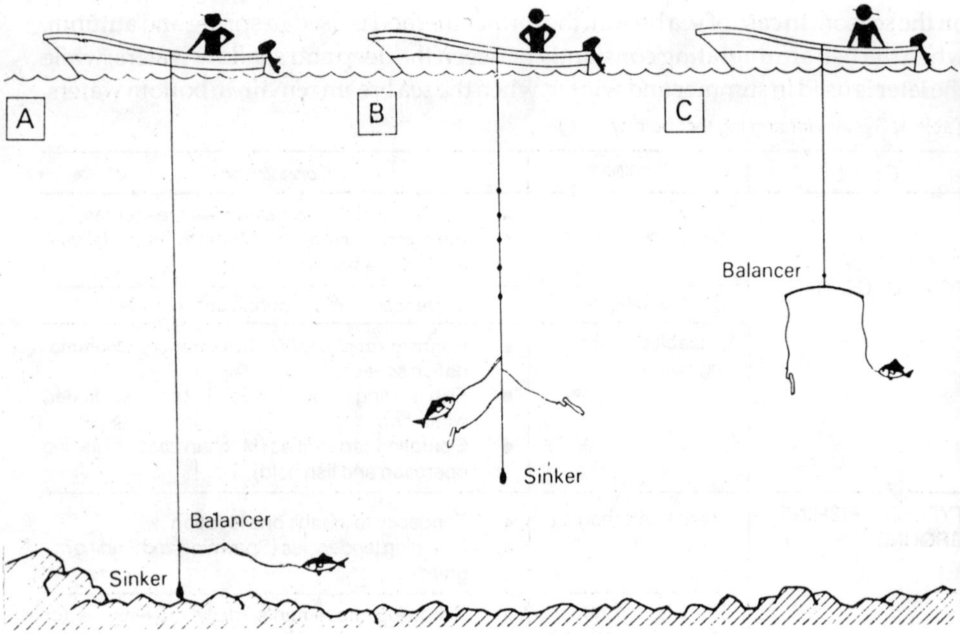

Fig. 14: Hand line fishing

A variety of baits, including live bait, fish pieces, paste bait and lures are used depending on the type of fish and the size of fish being caught, seasonal and sea conditions. Fishermen, at time also spread the baits in the water or use fish attracting lights, at night, to increase the fishing effectiveness.

Recently man-made artificial reefs are used in gathering fish for effective fishing.

Angling of Grunt (Parapristipoma trilinealum)

Grunt inhabit in coastal waters attain a size of 40 cm. in adult. As mentioned earlier, the most important point of the angling operation is to locate the habitat and the migrating depth of the fish to be caught, and then to place the hooks accurately in this range. In the case of bottom fishes with their strong tendency to stay in one place with regard to current and bottom topography, the success of the fishing operation depends largely on the fisherman's ability to locate the fishing ground, based on his knowledge of the preferred habitat and the movements of each fish species. Grunt inhabit open sea waters with a depth of 5 to 50 metres, tending to gather mainly in areas with considerable up-and-down variation in the rocks of the sea bottom. In summer, the grunt inhabit in rather shallow reef areas at a depth of 8 to 12 fathoms. At times of good biting conditions the grunt tend to gather along the edges and slightly above the rocky protrusions rising from the sea bottom. When biting is poor the grunt are said to distribute along the lower edges of the reefs.

The second most important decision involved is the time of fishing operation during the day. Fish do not feed all day long; instead, they have certain times of the day at which their feeding instinects are stimulated by such factors as tidal current or the amount of light penetrating the water. The fishermen should know at which time the fish of a given species bite and caught, such as around sunset and sunrise or moonrise and moonset, at the beginning of the tidal change etc., and should schedule their fishing accordingly, going out in the predown or at late afternoon.

The third factor is the choice fishing hook and bait. By trial and error, the proper hook and bait to fit a given fish's size and feeding habits is to be determined.

The final factor is the actual line handling skill of the fisherman. Each type of fish has a different way of taking the bait, and the fisherman should be sensitive enough to these differences to be able to respond properly.

In case of grunt angling, after the sinker has hit the bottom the fisherman immediately begins to pull the line up. When the hooks reach the mid-depth, the line is again lowered to the bottom and the process is repeated. Since the grunt is a fish which hooks easily, if the line suddenly becomes heavy as it is being raised, that means that a fish has been hooked and the line should be hauled in immediately.

The hooking technique of centrophid fish is, however, different from that of grunt. As the fish's teeth are situated not in the mouth but in the gullet, it feeds by starting at the end of the bait and tries to gradually swallow it. Therefore, the angling technique for centrophid fish is not to respond to the first gentle touch that one feel on the line, but wait for a harder second tug, which means that the bait has been swallowed and the fish is starting away. Only then one should pull the line rapidly.

Pole-and-Line

Using a pole to which is attached a line and fishing hook, is a method in which the pole is hand held and the hooked fish is brought up manually.

A 5-12 tonnes boat can conduct both bottom long-line fishing for sea bream and pole-and-line fishing for skipjack. Bottom long line fishing is done at the edges of the continental shelf (200 m. depth), whereas the fishing grounds for skipjack include the areas of 200 m. isobath and the edges of reef areas in the open sea. The boat for open sea fishing should have (i) good wave-resistance capacity and (ii) sleeping accommodations for the crews.

The fishing gear consist of a line and hook attached to a pole. Earlier, bamboo was used for these poles, but today fibreglass is replacing bamboo. The poles have a standard length of 4 to 4.5 metres and a diameter of about 3 cm. at the handle, and are tapered toward the end. The pole must have the right amount of flexibility to allow it to bend as the fish is being pulled in, and to accommodate different sizes of fish there are different poles prepared for 3, 4 and 5 kg. fish.

Two types of hooks are used, lure hooks and live baits hooks. Lure hooks are used for schools of fish with strong feeding behaviour. The hook, hidden by a feather fly, is jiggled back and forth along the water surface like a small fish jumping from the water. Because the hooked fish is raised immediately and released from the hook again by simply striking it against the deck of the boat, a hook without barb must be used. The line is made slightly shorter than the length of the pole, in order to allow the hooked fish to be swung easily through the air and into the boat.

Live bait hook is used for catching fairly large skipjack of over 3 kg. and for young tuna. Fishing operations should begin with lure hooks and then change over to bait hooks as the school's appetite wanes. Live anchovy or sardine of 6-10 cm. are attached to the hook and allowed to swim near the water surface until a fish takes it. The hooked fish is raised immediately and the fisherman catches it with his left arm and holds it as he releases the hook.

On locating a fish school, it is approached carefully from behind so as not to disturb its movement. After pulling up to the front of the school in a slow circling motion, the bait spreading starts and the fish are lured closer to the boat.

When the boat is positioned in a way that block, the front of the school, the engine is shut off and water is sprayed on surface of the sea from the side of the boat to attract the fish up close to the boat, at which point the fishing operation begins. A close watch is kept on the movement of the school and additional bait is spread if necessary.

As the number of fish passing around the boat and getting away begins to increase, the fishing operation is stopped, the engine started, and once again the school is approached carefully from behind and the fishing process repeated. A hungry school of fish is generally easy to bring up to the boat, while a school which is well fed will tend to scatter and get away.

Fig. 15: Pole and line fishing

Vertical Line

This fishing method utilises numerous fishing hooks on one line suspended vertically in the water, in order to increase the effectiveness of fishing. Vertical line angling can be divided into three main types:

(a) A multiple-hook long-line is suspended vertically in the water and moved in an up and down motion.

(b) A similar long-line is attached to a float, made of wood, plastic and allowed to drift with the current.

(c) A vertical long-line is attached to a float, but a sinker is attached to the end of the line and sunk to the sea bottom to hold the line in one place.

In all these above methods, the length of the line is set after determining the depth at which the fish are migrating. Either fish pieces, small shrimp are used as bait.

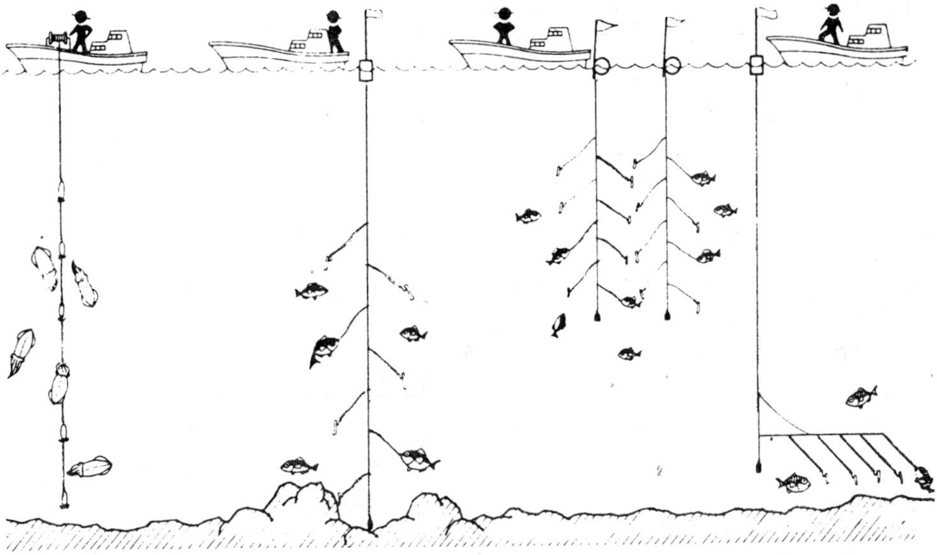

Fig. 16: Vertical line fishing

The fishing boats involved in vertical line fishery for bottom fishes are of two groups, e.g. (a) boats around 3 tonnes and (b) boats of 5 to 10 tonnes. The composition of fishing gear is the same for both small and large boats. The specifications of the fishing gear vary slightly with each species of fish. The differences are in the amount of bait and the method of attaching the bait and the fact that the number of branch line changes with the change in the migrating depth of the fish between day and night.

Stercolepis ishinage usually inhabits deep sea reef areas at a depth of over 200 metres, but during their spawning season in summer (May to June) they migrate to coastal reef areas with a depth of about 100 metres to lay their eggs.

Sebastes matsubari (grey rock cod) lives in schools in the crevices of rocky bottom areas having sharp up-and-down topography at a depth of 200 to 300 metres.

Fishing gear specifications for vertical line

Bottom fish with max. length in cm.	Fishing season	Fishing ground depth in m.		Bottom type	Main line		Trunk line		Leader line			Bait
		Night	Day		Thick-ness in #	Length in m.	Thick ness in #	Length in m.	Thick ness in #	Length in m.	No.	
Alfoncino 50	Through-out the year	150-250	350-450	Rocky reef	Super-tetron 30-35	450	Nylon mono-filamt 20-30	1.5	Nylon mono filament 20-30	1	20-30 (Night) 70 (Day)	Squid Anchovy Mackerel fillet
Bigeye 60	"	100-150	350	"	"	400	"	2	"	1	10-15 (Night) 20 (Day)	Anchovy Mackerel fillet
Centrophid fish 90	" (Inter-mitent)	50-150	—	"	Super tetron 40	150	Nylon filament 24-30	5–13	"	2-3	5-10	Squid
Gray rock cod 100	Winter Dec-Mar	—	500-700	"	"	700	"	1.5	"	1	20-30	Squid golden bait
Black Cod 130	Through-out the year	—	550-800	"	"	1000	Nylon mono-filament 50-60	3-5	Nylon mono-filament 40-50	1.5-2	3-5	Squid 30 cm. put on the hook whole

Ocycrius japonica (centrophid fish) live in the waters around plateau-shaped reefs at a depth of 70 to 90 metres forming dense schools slightly up current from the top edges of the plateaus.

Trolling

Trolling is the most active type of angling operation, and the vital points involved in the fishing process include: (a) the location and persuit of a school of fish; (b) setting the proper trolling depth for the fishing line and (c) the performance of the fishing hooks and bait.

Trolling fisheries may be divided into two major types.

Type I

This method is one in which the lines and hooks are towed horizontally near the water surface. The fish caught by this method are piscivorous species with a large migrating range, such as, middle to large sized tuna, marlin, skipjack, Spanish mackerel, yellowtail. In case of small fishing boats, one to three lines may be towed, but in middle to large sized boats may extend poles from either side of the boat to allow six to eight lines to be towed. Effective lures have been developed which attract the fish by the quick "swimming" motion they stimulate as they move through the water for this type of fishing. In commercial fishing, the boat size is 3 to 5 tonnes.

Type II

This fishing method aimed at catching middle to bottom depth migrators, such

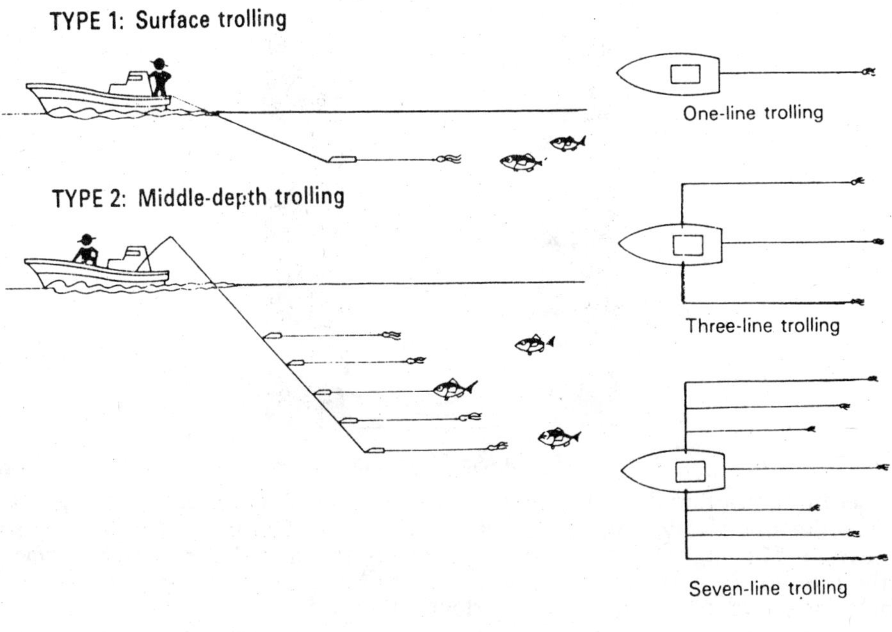

Fig. 17: Trolling

as sea bream, yellowtail, trout and flat fish. The line is equipped with devices like sinkers, underwater plates and floats that cause the line to troll in the middle to bottom water depths. The line is fitted with multiple lures towed through the water and the design of fishing gear is similar to that of vertical line and can be considered a type of long-line fishery.

Upon reaching the fishing ground and locating the fish school, the boat should approach the school in a way that will not disturb their movement and begin the trolling process, sweeping in front of the school at a speed of 3 or 4 knots. During the process three factors must be considered, e.g. the direction of the sun, the movement of the boat and the movement of the fish school.

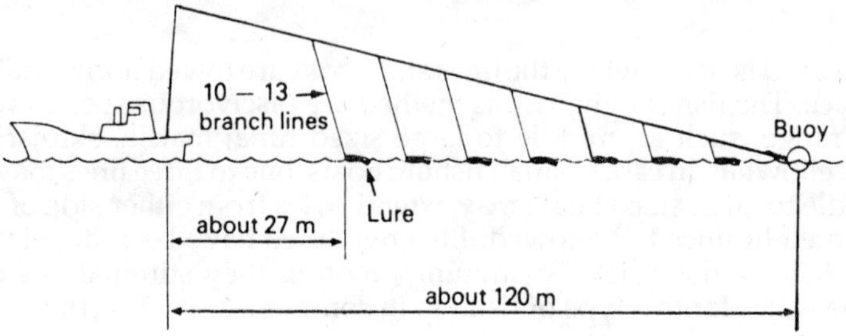

Fig. 18: Surface trolling

Rocky reef areas around island tend to be feeding grounds for migrating fishes and in such fishing ground, it is best to search for the schools of fish by starting near the shore and following isobath lines one after another out into the deeper waters.

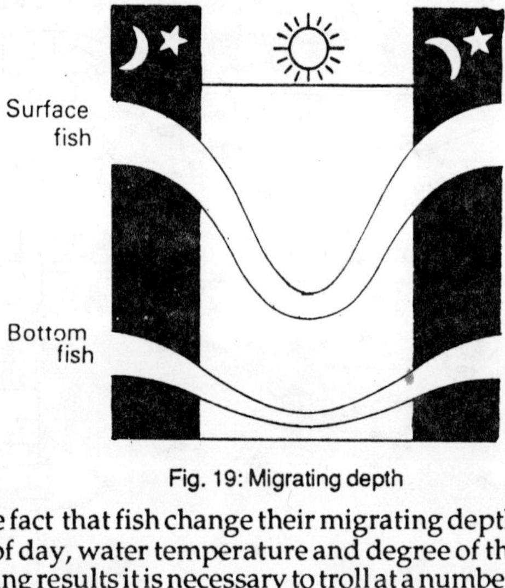

Fig. 19: Migrating depth

Because of the fact that fish change their migrating depth depending on factors, such as, the time of day, water temperature and degree of their appetite, in order to achieve good fishing results it is necessary to troll at a number of depths from surface to the middle. To make the line and hooks troll at desired depth, devices, like submarine board or a surface-planing device are used.

By adjusting the angle of submarine board in relation to the surface, the depth at which the line and hooks will be towed is controlled. The surface-planing device on the other hand is designed so that its smooth surface receives the waves and rides up over them easily. In addition to its function of riding on the surface of the water, this device also function as a vibrator to give a jerking motion to the baited hooks behind it of the sort that will stimulate the feeding instincts of fish.

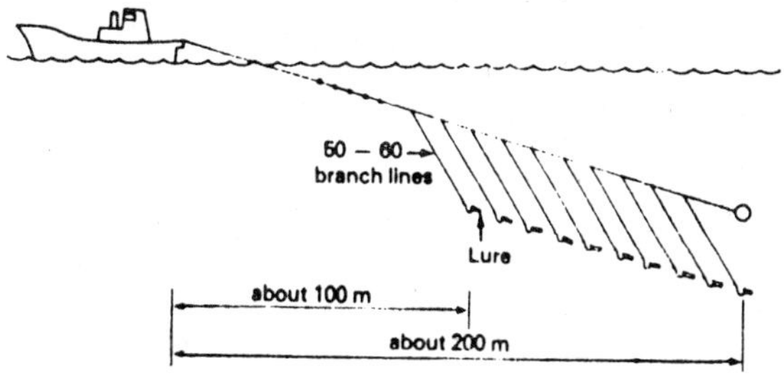

Fig. 20: Mid-water trolling

Both live bait, such as squid pieces, and lures, such as, feather fly are used in trolling. Commercially made and fishermen's own hand made lures are of variety of shapes and materials. In case of both live bait and lures, it is extremely important that the bait be one that appeals strongly to the fish and stimulates its feeding instincts by (a) moving through the water with a quick "swimming" motion similar to that of a real fish; (b) that it provides a strong visual stimulus by means of bright colours and intermittent flashings of reflected light and (c) that it has a slight sound as it "swims" like that of a real group of migrating fish.

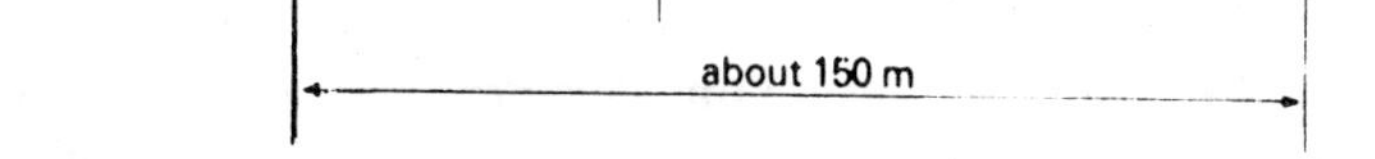

Fig. 21: Bottom trolling

Longline Fishing Methods

The basic structure of a longline (Fig. 22) consists of a long main line (trunk line) to which branch lines are attached at regular intervals, each of which ends with a leader or snood to which hook and bait are attached. After the line has been set in the water, marker floats and flags (marker lights) are attached either to the float line at either end of the main line or marker float lines attached to the mid-section of the main line, in order to mark the position of the line. In case of bottom longline anchors and anchor ropes (attached to either end of the trunk line) and sinkers (attached to the mid-section of the trunk line) are necessary in addition to the float and marker lines stated above.

Longline can be divided into two basic types: the drift longline which uses floats to hang the main line (trunk line) at desired water depth, and the bottom longline which is set in such a way that the trunk line and branch lines just come in contact with sea bottom. In recent years, a new variation of bottom longline, known as vertical-line type bottom longline in which trunk line is floated about 10 metres from the sea bottom to allow the branch lines to hang down vertically to the bottom.

Drift longline is used for catching fish that migrate in the surface waters or slightly below, with the main catch consisting of tuna, skipjack, marlin, shark, salmon, trout, seabass, black rock-fish, half-beak and others. This category also includes mid-depth type long-lines used in catching Alaska pollack, mackerel and yellowtail.

Bottom longline is most popular and is used in all the coastal regions to catch a wide variety of fish. The main catch includes fishes of the cod family, flounder, flatfish, sea bream, tilefish, mackerel, drum croaker, hairtail etc.

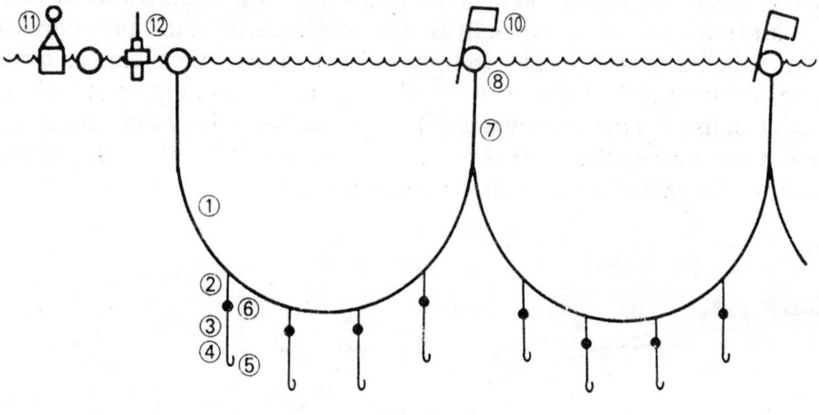

Fig. 22: Drift longline

(1) Trunk line, (2) Branch line, (3) Sekiyama, (4) Snood, (5) Hook, (6) Swievel, (7) Buoy line, (8) Float, (9) Anti-pressure float, (10) Marker flag, (11) Marker light, (12) Radio buoy and (13) Anchor.

Vertical line type bottom longline is a improvement on the bottom long-line, having the advantage of being able to be set even in areas with rocky reef bottom topography. It also combines the advantages of being capable of distributing a large number of hooks over a large area of the sea bottom and the ability to catch not only bottom feeding fish but also fish that migrate and feed slightly above the sea bottom. The main catch includes bigeye, deep sea bass, parrotfish, centrophid fish and others.

As a fishing method, longline has the advantage of being able to extend over a large water area. Pelagic tuna longlines may be lengths between 50 to 60 km. or even up to 100 km. in length, while in coastal waters fishermen operate longlines of 10 km. length. In order to facilitate the fishing operation, the trunk line is usually divided into sections of about 500 metres, called one "basket", which is kept in basket-shaped containter along with its branch lines, snoods and hooks. Each of the "basket" containers with their fishing gear, including trunk line, branch lines, snoods and hooks and bait are prepared in the same way so that they can be easily connected one to another in the line casting operation without any problems. When the line is put back in the basket after use, it is done in a way that will ensure convenient use in the next line casting with the trunk line stacked in a "flower petal" pattern neatly, and the branch lines piled separately near the rim of the basket so as not to become entangled with each other, and the hooks are stuck lightly into the rim of the basket at regular intervals. To make the end of the trunk line easily identifiable, it is pulled out of the basket slightly and left hanging out.

Vinylon are used for the trunk lines; while branch lines are usually made of nylon gut. The thickness of the lines is determined by considering the size of the fish to be caught, depth of water in which the line will be used and the overall length of line to be used in the fishing operation.

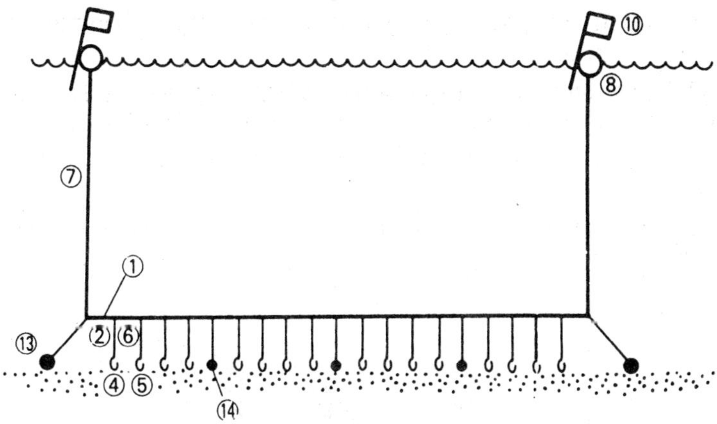

Fig. 23: Bottom longline

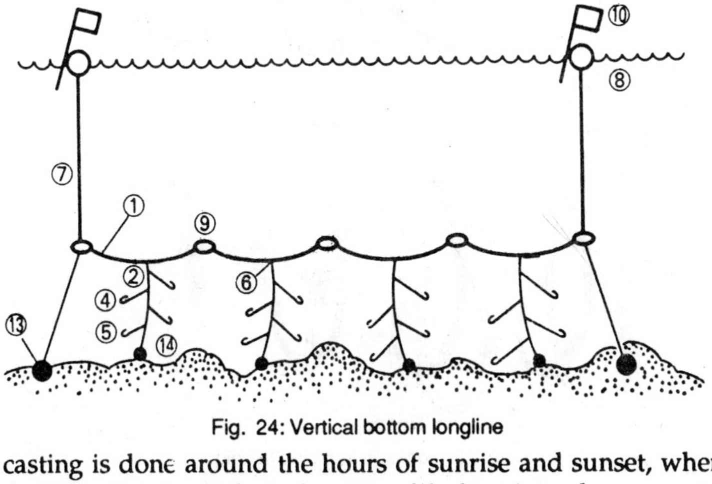

Fig. 24: Vertical bottom longline

Line casting is done around the hours of sunrise and sunset, when the fish normally feed. The line is cast from the stern of the boat in such a way as to have the line lie perpendicular to the tidal current flow, for migrating fish coming in contact with the hooks and also prevents the branch lines from becoming entangled with the trunk line, a condition which often occurs when the trunkline is cast, parallel to the current flow. In reef areas the trunkline is cast in a zig-zag pattern, although the line is usually cast in a straightline in case of both drift longline and bottom longline.

The hauling operation of the line may be, (i) immediately after casting, (ii) one or two hours after casting or (iii) in the morning after leaving the line out all night, depending on the region and the type of fish being caught.

Hook for Longline Fishing

The hook used in longline is one with a curved-in point with most having a relatively narrow gap. This type of hook will set itself in the fish's mouth region and not cause injury to other parts of the fish body, and is capable of holding the catch for a long time:

(a) For sea bream like fishes a round-shaped hook with an extreme curve-in at the point is used.

(b) For globe fish sting ray, hairtail and eel a hook with a long shank is used to allow for easy dislodging of the hook after the catch is hauled in.

(c) For flatfish, mackerel and yellowtail type fishes an angular hook is usually used.

(d) For large sized fish like tuna, shark, marlin a large hook with an especially large barb is used.

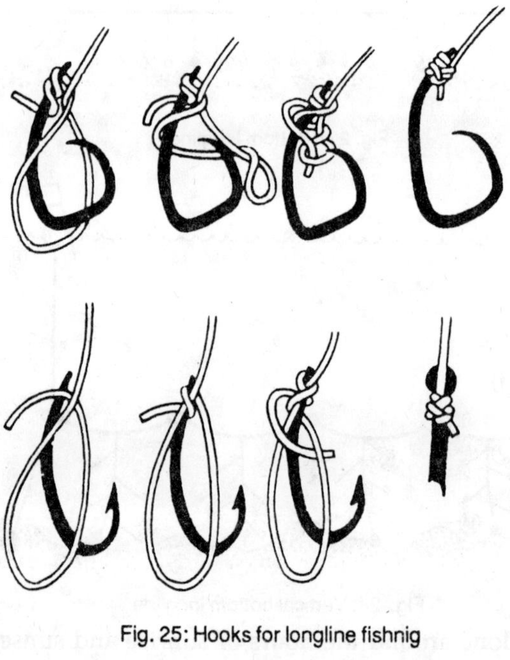

Fig. 25: Hooks for longline fishnig

Tuna Drift Longline

The fishing boats used for tuna longline fishery vary from 5 to 400 tonnes. There is also variation in the thickness of trunk lines and branch lines, the number of branch lines, the dimensions of the hooks and the number of baskets used.

In small scale coastal fishery, 50-60 metre length of line are connected together to form one trunk line unit (one basket) of 300-350 metres. The number and length of branch lines attached to one "basket" of trunk line varies on the type of fish to be caught and the condition of fishing grounds. For bluefin tuna, yellowfin tuna and bigeye tuna there is a standard to have 4 to 5 branch lines of 15 to 20 metres in length, while Albacore usually calls for 8 to 12 branch lines of 15-20 metres.

Since a line hauler is used, it is necessary to use line material with good flexibility and good strength under tension. In tuna longline vinylon is the most common material used. The usual trunk line thickness is from 4 to 6.5 mm. with branch lines being the same or slightly thinner.

Branch lines are made up of (1) branch line, (2) "sekiyama" and (3) snood and hook. Wire is used in the snood for reducing visibility and the "sekiyama" (a steel wire covered with coiled cotton thread) in the mid-section serves to facilitate handling during the line hauling process.

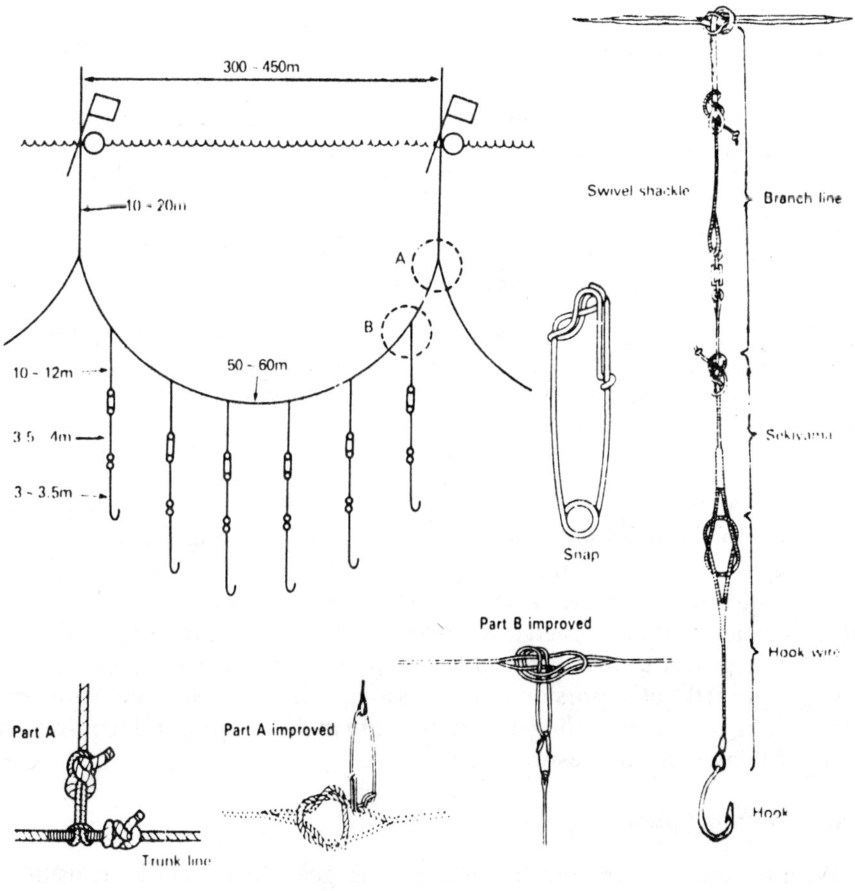

Fig. 26: Tuna longline fishing gear

For effective line operation, it is important that the connecting and disconnecting of the trunk lines, branch lines and float and marker lines can be carried out easily and surely. Traditionally this task relied on the skill of the fisherman's rope work, but recently there has been the use of snaps in tuna long line fishery. The invention of these strong and easy-attaching snaps has made it possible to disconnect the branch lines, while the trunk line is being reeled in continuously during the hauling operation.

Nylon Gut Type Tuna Longline

The greatest advantages of nylon gut are that it is thin, strong, light and also transparent; and when used with live anchovy bait, the bait appears to be swimming in almost a natural state, thus improving the catch effectiveness. But the problems remain with harder handling than polyvinyl twine due to stretchability of nylon monofilaments and its breaking in the connecting points.

The following improvements have been made to overcome the above difficulties:

(i) Thickness of the nylon gut used in the trunk and branch lines were increased to # 90.

(ii) The wire snood was eliminated and the hook attached directly to the nylon gut, so as not to inhibit the swimming action of the live anchovy and the number of windings used in attaching it to the hook was increased to prevent breaking.

(iii) The trunk line and branch lines are hauled in together and the swivel used in the attachment was improved to prevent entangling of the lines.

The fishing grounds are located at 112-160 km. from the coast at a depth of 300 to 1000 metres; and the fishing season is roughly 8 months from April to November. The catch is made up of about 70 per cent yellowfin tuna, followed by marlin family, Albacore, dorado and shark. 5 to 10 tonnes class boats are engaged in this type of fishing.

At night a group of boats leave for fishing ground. On reaching fishing ground the boats separate at 3-4 km. intervals and turn off their engines and drift. Line casting begins at dawn, when the fish begin their feeding activities. Fishing is done for three hour from 6 to 9 A.M. The line is cast across the current flow, from the landward side and the line is left to drfit in the water for six hours before the hauling operation begins. Fifteen to thirty "baskets" are used in one fishing operation. The hauling begin at the offshore side and works toward land to save unnecessary travel by the fishing boat. The hauling operation begin in the evening and requires about 6 hours. At dawn the line casting begins again.

Vertical Bottom Longline

With an ordinary bottom longline fishing gear, there will be considerable damage to the trunk line and branch lines when it is used in areas with rocky reef bottoms, and the trouble often arose from the hooks being caught on the rocks. The

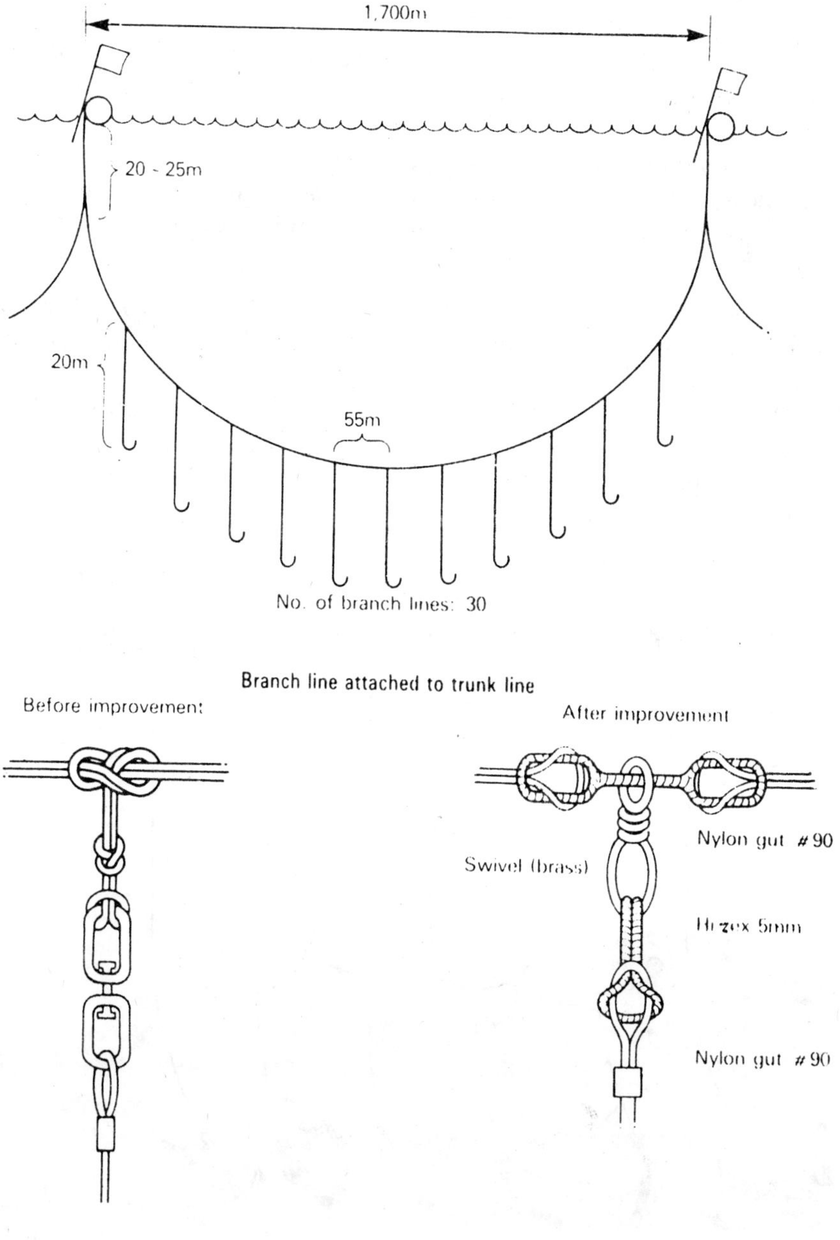

Fig. 27: Nylon gut type tuna long line fishing gear

improved fishing gear greatly decreases such problems, through the following adaptations:

(i) Pressure-resistant type floats made of plastics attached to the trunk line cause it to remain suspended in the water above the sea floor and causes the branch lines to hang down vertically, preventing the hooks to come in contact with the rocks on the bottom.

(ii) Iron rod of 3/8" hickness, cut into 180-200 cm. length and the upper 1/6th of the rod bent at 30° angle are used as sinkers to prevent becoming caught on the rocks.

(iii) A slightly thinner line to be used at the point where the sinker is attached to the end of the branch line so that if, by chance, the sinker is caught on a rock pulling strongly on the line will cause it to break off at the point of attachment.

Since the fish to be caught include reef inhabiting fish, the meat of mackerel or saury pike is usually used as bait, with salted or frozen meat being put on the hooks before leaving port. The boats leave for fishing in the early morning so as to arrive at the fishing grounds by sunrise, the time when the fish begin their feeding activities.

On arrival the desired fishing depth is determined with the help of a fish finder and the line is cast along the topographical line of the chosen depth or on the

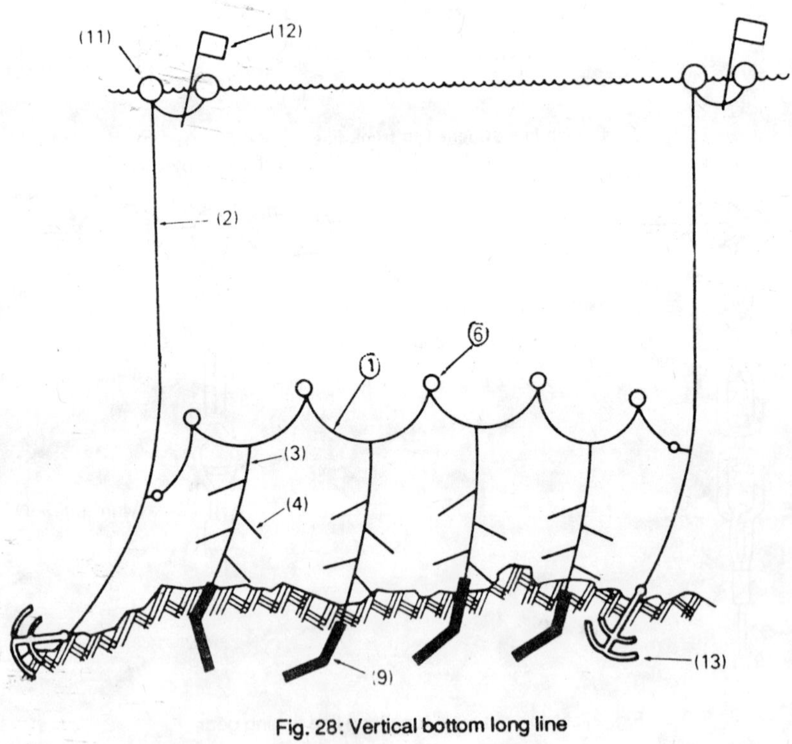

Fig. 28: Vertical bottom long line

up-current side of a reef. The hauling operation begin after $1^1/_2$ to 2 hours after the line is cast. It is desirable that the fishing gear have an appropriate sinking speed in relation to the current speed for casting the gear at the specific points on the sea bottom as fishing grounds. This is done by adjusting the material and number of pressure-resistant floats and weight and number of sinkers to be used.

Bottom Longline

The fishing season for tilefish is 9 months from September to May, with the fishing grounds lying three nautical miles from the shore.

Before leaving the port for fishing 60-70 baskets of lines is prepared and the hooks baited. The bridge is made large enough to enable two fishermen to spend one night at sea.

During line casting operation, one person drive the boat, while the other casts the line from the rear of the boat. In the hauling operation, one person will haul the line from the front of the boat, while the other person takes the fish caught from the hooks, sorts and stores the catch.

Salted squid and mackerel pieces are used as baits.

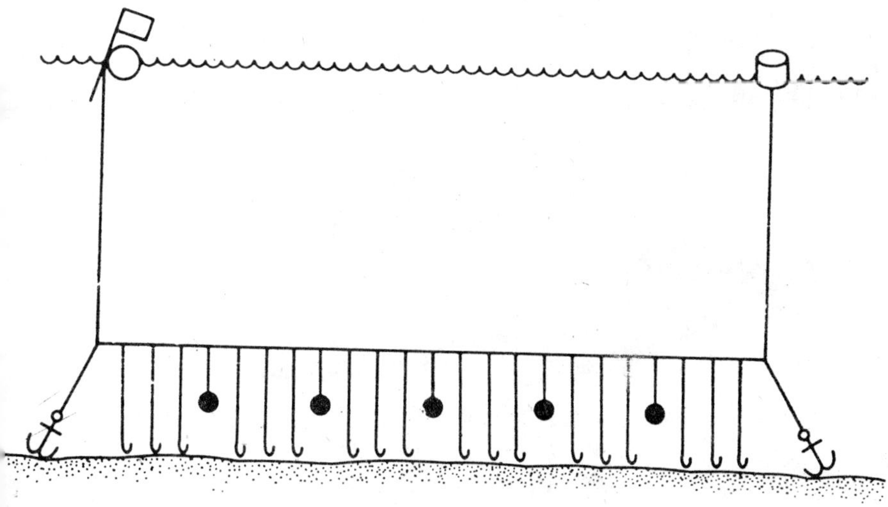

Fig. 29: Bottom long line

4

TRAWLERS

Beach Seine

A semi-circular net is spread in a shallow area near the beach. Ropes connected to the ends of the net are pulled from the land to catch fish living near the beach. In drawing the net, the wing nets are at first fully extended, then brought closer together as the net is brought closer to the shore.

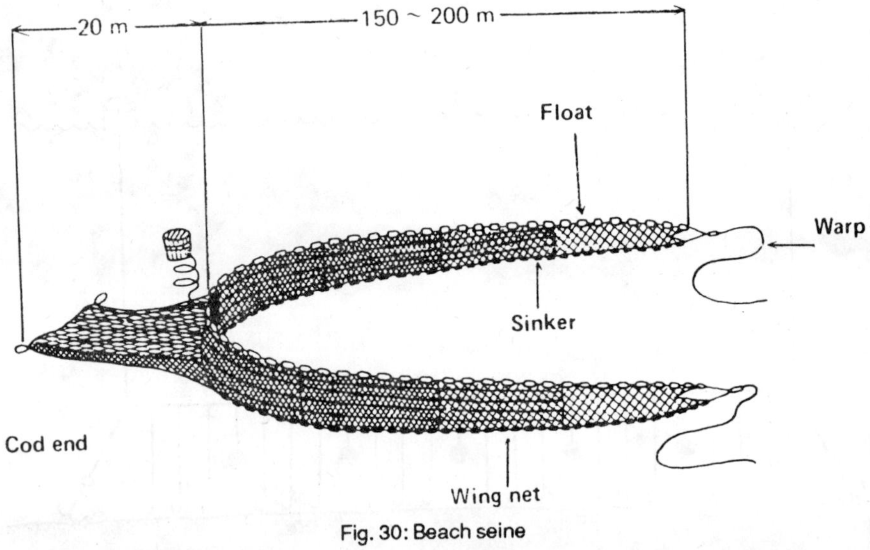

Fig. 30: Beach seine

Trawler and Trawl Net

Trawler means, any vessel that pulls a bag net through the water and the bag net thus pulled is known as trawl net. Granton trawl nets or its extended version continue to represent a large proportion of the gears currently in use in Britain. Modified Granton also remain in demand in many other parts of the world.

There are many types of trawlers varying in size from open boats powered by outboard engines to huge factory ships, which is designed to fish in distant waters

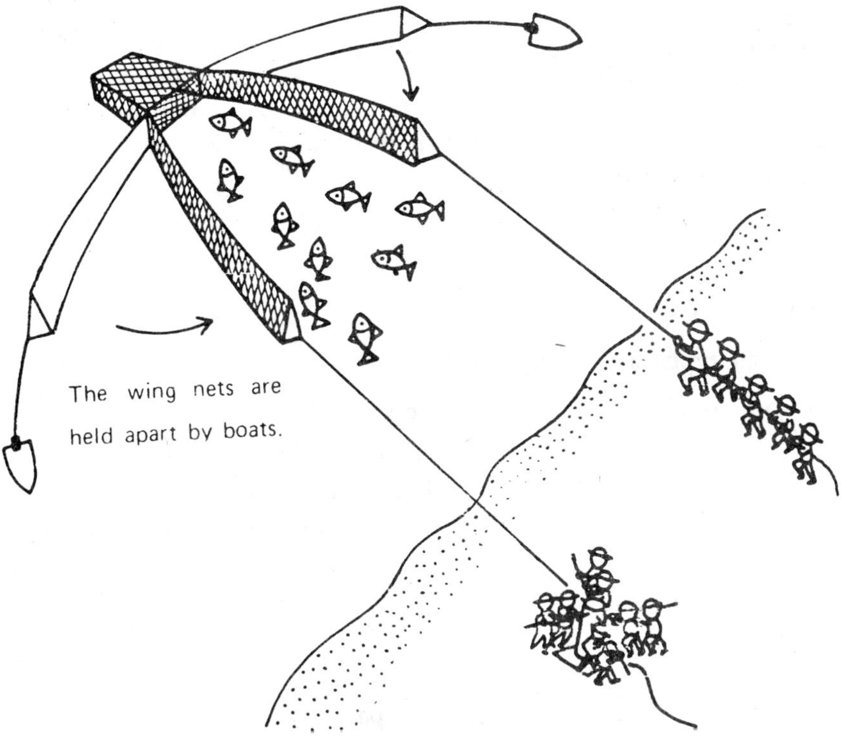

Fig. 31: Beach seine in operation

Trwaler is, therefore, a very general term and for precision, the type of trawling engaged in and the size of the vessel is important.

The simplest type of trawl consists of a bag net without an opening apparatus at the mouth. This type of trawl can be further divided into (a) trawls towed by one boat; (b) trawls towed by tying tow lines to outstretched poles at the bow and stern, while using the tidal currents or wind for power; and (c) trawls towed by two boats. These are long-established fishing methods, but they have gradually been replaced by the more effective beam trawls, dredge nets and otter trawls that have been developed with the increased motorization of fishing boats. In Japan, however, two-boat trawls are still in operation in off shore areas.

Trawler Requirements

Two basic requirements are common to all trawlers. One is the need for towing power and the other is the need for a winch or mechanical hauling system. In order to have good towing power, a trawler should have a reasonable draft or displacement, and should have a large slow-turning propeller. With the high r.p.m. engine, big reduction gear is a must, so as to give the propeller a speed of not more than 400 r.p.m. at highest engine speed in order to save wastage of power through propeller cavitation. On larger trawlers the propeller speeds are much less. A medium or slow-speed engine with a smaller reduction gear is preferable.

Though there is a limited advantage of more towing power with propeller nozzles, yet the effect of the extra thrust is good on stern trawlers, but on side trawlers which often must use a lot of helm to counter wind and tide, the effect is minimal. The indirect benefit of a nozzle is that it provides more protection to the propeller during shooting the gear; but a net can foul a nozzle propeller just as easily as it can foul an open one.

The winch or hauling mechanism is the next important requirement for a trawler.

Types of Trawlers

The main types of trawl vessels are:

(a) Otter trawlers	(i) Side trawlers
	(ii) Stern trawlers
	(iii) Double rig trawlers
	Side or stern trawlers, may, engage in bottom trawling, semi-pelagic trawling or mid-water trawling.
(b) Pair trawlers	(i) Bull trawlers
	(ii) Pair seiners
	(iii) Cod pair trawlers
	(iv) Mid-water pair trawlers
(c) Beam trawlers	(i) Single rig
	(ii) Double rig
(d) Others	(i) Scallop dredgers
	(ii) Danish seiners.

Otter Trawlers

The use of otter boards to open the trawl net mouth horizontally in preference to a beam, was introduced over hundred years ago. Otter trawling developed rapidly after 1920 and by 1950 it was the single most productive method of fishing. In the last 20 years there have been further significant developments in otter trawling. Though otter trawlers are classified into three main categories: side trawlers, stern trawlers, and double rig trawlers; there are actually a large number of variations in the gear used. They are shrimp trawls, prawn trawls, combination trawls, wing trawls, bobbin trawls, herring trawls, semi-pelagic trawls, and midwater trawls. Most trawlers use one type of net for the greater part if not, for all the year.

The smallest otter trawlers have gallows, gantries or derricks to handle the heavy otter boards. The net hauling system varies greatly depending on the size of the vessel and the type of the trawl used. Light wing trawls or mid-water trawls may be hauled in by power blocks or net drums. Heavy bobbin trawls may be lifted aboard with quarter ropes or gilson wires.

Most trawl gallows are of the inverted "U" shape, but there are a variety of designs. They are tripod, bipod, single pillar and gantry arrangements. For selecting trawl gallows, the arrangement should be high enough to accommodate the largest boards to be used, strong enough to take the maximum anticipated load and will cause minimum interference to other activities on board.

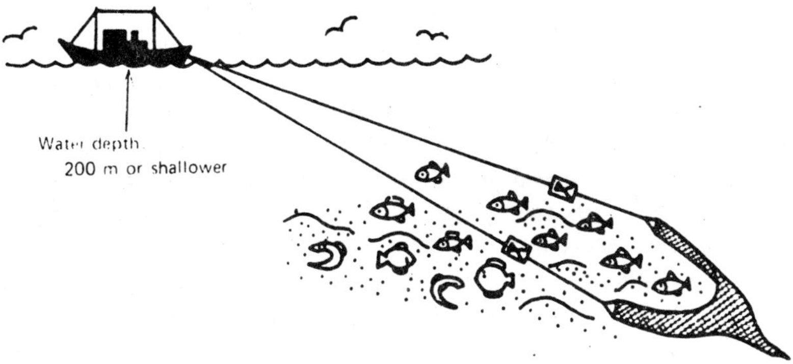

Fig. 32: Otter trawler

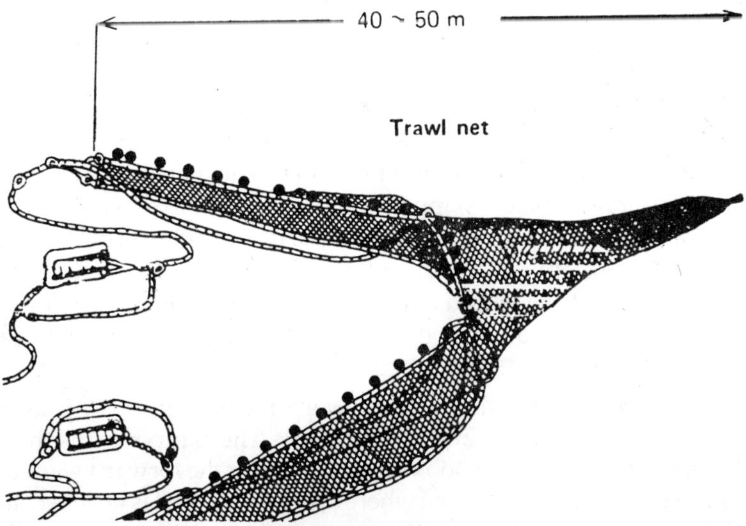

Fig. 33: Otter trawl with otter board

Trawl warps must run out from the winch to the gallows blocks or gantry blocks. The position of the warp leads on deck and the "run" of the wires should not cause annoying delays in shooting and hauling and result in excessive wear on the gear or unnecessary loads on the winch. The warp must lead in at right angles to the winch drums. From the first pair of deck leads, the warps are led to the gallows blocks by an arrangement suitable to the type of vessel and method of operation. On side trawlers they usually run to the rail and down the side deck; while in stern trawlers they may lead directly to a roller below the gantry.

Side Trawlers

The distinguishing feature of all side trawlers is the two steel gallow frames, one placed aft and the other forward on the starboard side. The starboard otter door is hauled up to the forward gallows and the port otter door is hauled up to the aft gallows block. There are slight differences in the shooting and hauling operations between large deep sea side trawlers and small "light trawl" side trawlers.

The deep sea trawler lies with its starboard side to the wind and the cod end is put over the side. As it drifts away from the boat, the bag and body of the net are paid out. While the headline is held clear, the bobbins and ground rope are hoisted up by their quarter ropes and dropped over the side. The winch breaks are released and the dan lenos and bridles or cables are allowed to slide out after the net. At this point the skipper gives the vessel a turn ahead then stops as the last of the cable runs out.

The end of the cable is attached to a "keep" which is a figure of 8 shackle with one end much wider than the other. All of the cable has been running out through a "Kelly's eye", which is also a figure of 8 shackle but of different design. The Kelly's eye is attached to the back strap of the otter board. The keep jams there as it is too big to pass through and the weight of the cables is taken on the otter board which is hanging in a chain in the gallows. Between the keep and the warp end is a length of wire or pennant. It hangs loosely over the otter board. The warp end shackle is attached to the trawl door bracket, the winch pulls in the warp until it takes the weight of the boards. The chains are unhooked and the gear is ready to be shot.

The skipper steams the trawler ahead and slightly to starboard. Signals are given to the men at the winch to release the breaks before turning slightly to starboard and the doors are dropped into the water. The men on winch maintain some tension on the brakes and the skipper watches carefully to make sure that both the otter boards are spreading correctly. The warp wires are now set out in a straight course with the trawler steaming at nearly full speed. The man on winch watches the warp marks and calls out the length at each 25 or 50 fathoms. As the final mark approaches the trawler slows down and the winch brakes are tightened when both marks are level at the stern.

Before towing commences, the forward warp must be pulled up to the aft gallows and both warps fastened in a towing block. The skipper turns the ship's head to port, a messenger wire and hook are thrown over the forward warp and by means of this it is winched into the stern where the towing block is fastened around both wires. The vessel now resumes its proper course and the towing speed is adjusted. The warps should be spreading outwards slightly and leading down from the towing block.

A normal tow with an otter trawl lasts for 2 to 4 hours. When the time comes to haul, the towing speed is reduced, the wires are released from the towing block and the winch is started. While the warps are coming in the boat is still going slow ahead and possibly slightly to starboard, the skipper being careful to ensure that when the net comes up, the starboard side will be to windward so that the gear will be up-wind of the vessel.

As soon as the otter boards are up to the gallows the winch is stopped, the holding chains are hooked on and the warp slacked back to let the chains take the weight of the doors. The warp end shackle is released from the bracket of the door and hauling recommences. The warp pulls in the pennant wire followed by the cables which are drawn in through the Kelly's eye. When all the cables are in up to the dan leno on the trawl net wing, the winch is stopped and net hauling starts.

The heavy bobbins are lifted aboard using the quarter ropes. The crew then pull in the body of the net, while one man winches in the cod end lazy line. With net aboard and the bag of fish alongside, the gilson hook is brought down and the cod end lifted aboard. As the bag swings over the pounds, the cod end rope is released and the catch spills out. The cod end is tied again and the gear may then be set again for another tow.

With small side trawlers, the net and bridles are paid out with the vessel going astern rather than waiting for the boat to drift downwind. The otter doors and warps are shot much as on a large trawler. Instead of using a towing block, many small trawlers simply put the foward wire into an extra open block on the aft gallows block, and both wires are towed from that point.

The hauling procedure is similar to that for the larger side trawlers. Some small trawlers use rope cables and bridles which may be bent on to winch barrel instead of winched on to the main drums through a Kelly's eye. Light trawls may be pulled in by hand or hauled on to a net drum or in through a power block.

Stern Trawlers

Fishing over the stern can be a very efficient way of trawling if properly designed. Mediterranean trawlers practised stern fishing for many years, but on deep sea trawlers the method did not come to widespread use until the 1960's. It was particularly successful on the large factory trawlers and freezer trawlers which could accommodate a stern ramp and a fish deck below the main gear deck.

The largest stern trawlers are able to have their trawl net wings laid out full length on the working deck. This is useful and appropriate on a deep sea trawler using heavy bobbin rigged nets. Small stern trawlers or stern trawlers using light nets have no such requirement.

Setting and hauling the gear from a stern trawler is an easier or simpler operation than from a side trawler.

The main points to be decided while designing a stern trawler layout are the positions of the bridge, engine room, and winch, and whether the vessel will have a stern ramp.

On most stern trawlers the bridge is placed forward. If the forward bridge is adopted, then it is to be absolutely sure that the skipper will have clear vision of the whole deck area aft of the bridge. Many stern trawlers are now being built with the wheel house aft of midships or just about midships. On stern trawlers of less than 60 feet (18 metres) overall length deck space is so limited that one has little option to put the wheel house forward.

Generally a fish hold is positioned as centrally as possible so the catch will not alter the vessel's trim unduly. On conventional boats with the bridge aft, the engine room is situated aft of the hold. But this can create difficulties in stern trawler. With the engine room placed forward, the trawler can work quite well if the vessel is soundly constructed and well maintained.

Most of the early stern trawlers had their winch positioned just aft of the forward-sited wheel house. On later vessels there is a preference for split winches of hydraulic or electric drive rather than single 2-drum winch, to have the warp go straight from the winch drum to the gallows or gantry avoiding too many right angle bend.

All larger stern trawlers over 30 metres in length have stern ramps up on which the net and cod end with full of fish are hauled in one easy pull. Without a ramp the cod end with fish must be lifted vertically aboard and if the catch is large, the bag must be "split" or brought aboard in two or three separate lifts. Ramps have been fitted successfully to trawlers of 21-24 metres or less.

There are three chief ways used to haul in the net on a stern trawler. One is to winch the wings aboard until the whole footrope and groundrope are in, and then to "fleet" the net in using beckets and gilson wire until the codend is reached which requires a stronger gilson wire. A second and popular method is to use a net drum. The wings and part of the bag are wound on to the drum which may be situated at any convenient point on the working deck. The third method is suitable on for small trawlers using light gear (without bobbins) is to use a power block. Power blocks of 24 inch diameter are usually adequate for stern trawlers under 100 gross tonnes, using wing trawls or midwater trawls.

The largest stern trawlers have a fish deck or factory deck below the working deck. A large hydraulically operated hatch opens just behind the cod end on the after deck and fish pour down into pounds in the fish deck below.

A good powerful deck wash and washing trough or pond should be fitted close to the fish hold hatch. The clean fish is transferred below by basket or by chute.

Crew shelter is important when working with the fish catch. It is necessary to provide maximum shade for the fish. On many large trawlers today, ice is sprinkled over large catches to keep the fish cool until they are processed or washed and packed.

Since most stern trawlers tow one warp from each quarter, the gear has that much initial "spread" before the wires enter the water. Thus for any given warp length a stern trawler should have better otter door spread than a side trawler. The stern trawler also enjoys the benefit of more direct thrust from its propeller than the side trawler. Stern trawlers using single towing block in difficult grounds have their wheelhouses placed forward and have a block or hook slung from the mast or derrick, in which the warps are held when towing.

Shrimp Trawlers

To catch shrimp only, the trawlers have been evolved on a particular design. The double-rigged "Gulf of Mexico" type first used in 1955, is the best known. The features of the shrimp rig have been developed to suit a particular type of low-headline net with long otter boards attached close to the wing ends. One such net is pulled from each of the side booms, which are extended during the fishing operation. A small try-net is sometimes also operated over the stern.

Shrimpers are interested in shrimp only and the deck layout and hold installations reflect the particular requirements of heading and freezing these high priced crustaceans. Since many shrimp trawlers operate far from port and for trips of several weeks duration, large fuel and water capacities are essential. The crews carried tend to be small and crew accommodation is usually above decks.

Shrimp trawl winches have parallel drums. Many are fitted with a third drum for the try-net wire. From the winch, the warp travels to the main mast or bipod and from there to the trawl through a "big boy" towing, block. Twin wires from each otter board are spliced into a single wire some 50 or 100 metres from the doors.

After the nets are pulled up to the booms, the cod ends only are drawn in to the vessel's side and hoisted aboard. This is done by means of a lazy line. The gear can then be re-set again very quickly. When towing, one warp is kept about 50 metres longer than the other so one trawl is always ahead of the other.

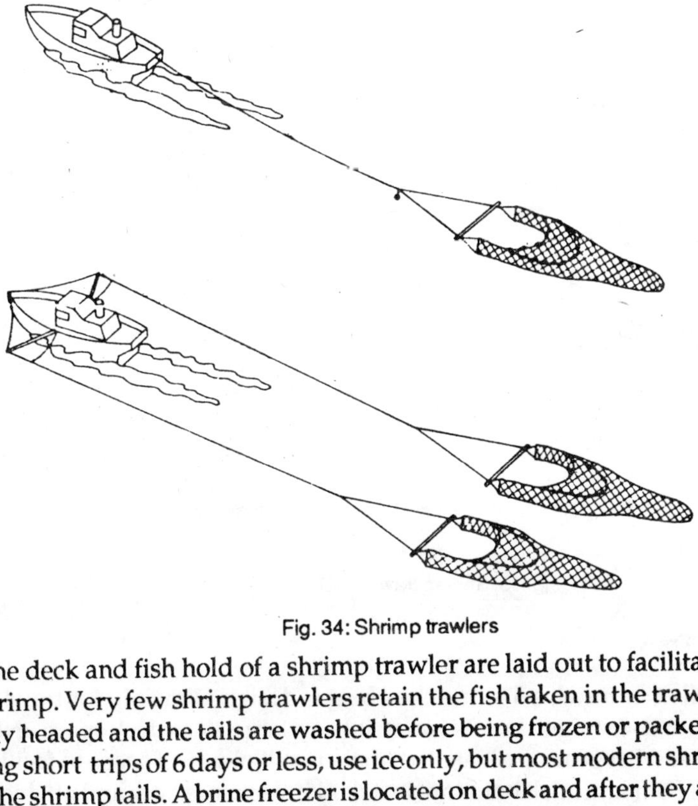

Fig. 34: Shrimp trawlers

The deck and fish hold of a shrimp trawler are laid out to facilitate processing the shrimp. Very few shrimp trawlers retain the fish taken in the trawl. Shrimp are usually headed and the tails are washed before being frozen or packed in ice. Boats making short trips of 6 days or less, use ice only, but most modern shrimpers freeze all of the shrimp tails. A brine freezer is located on deck and after they are frozen, the shrimp are packed in boxes in the refrigerated fish hold.

Many shrimp trawlers in Africa, Asia and India are small single-rig vessels. These boats work only one net. They usually operate close to port and use only ice to preserve the catch. These vessels are cheap to build and operate since they do not require large engines or refrigeration equipment. Many do not have even use a proper trawl winch. Two capstan heads driven by chain off the main engine are used to haul the warps in. In this case the warps are made of rope.

These small shrimp trawlers operate over the stern but they do not require gantries or gallows. A towing roller on each quarter is sufficient to hold the warp, and some boats do not even use that. The otter boards are small and light enough to be lifted aboard by hand.

There is no need for radical changes in designing such a traditional type of shrimp trawler, as the operation is simple and no sophisticated gear is required. The capstan or winch installation should be made as simple and strong as possible with the minimum of power loss from having too many connections or bends in the drive arrangement. The fish hold should be insulated on the sides, bulkhead and deck to prevent heat loss. A canvas awning should provide shade over the hold and working deck.

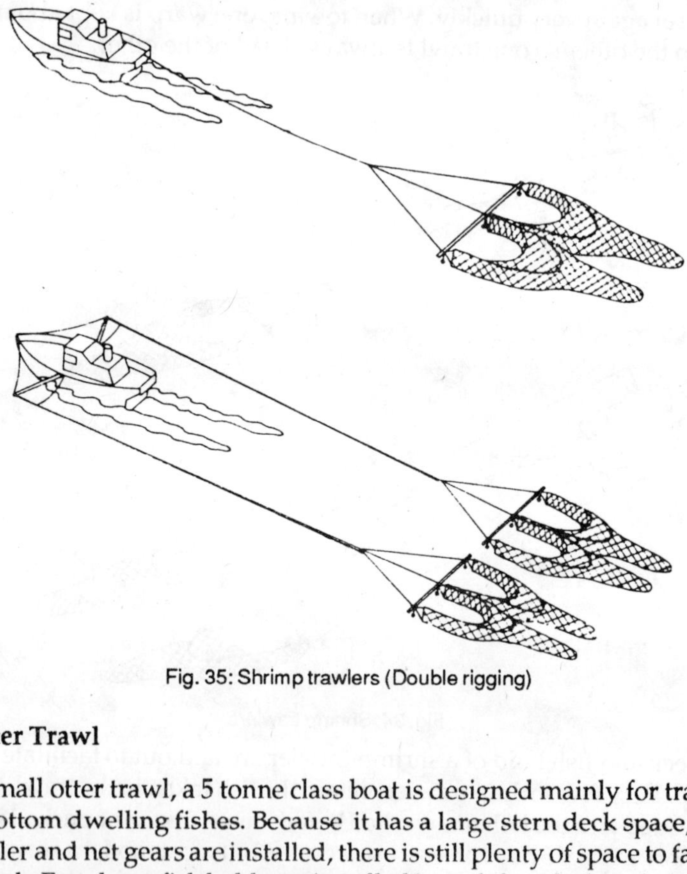

Fig. 35: Shrimp trawlers (Double rigging)

Small Otter Trawl

The small otter trawl, a 5 tonne class boat is designed mainly for trawl fishery to catch bottom dwelling fishes. Because it has a large stern deck space, even after the net roller and net gears are installed, there is still plenty of space to facilitate the fishing work. Four large fish holds are installed in each fore deck and aft deck. Keel-type hull bottom increases the boat's stability during motion and serves to prevent drifting by wind and rolling. It features a large bridge. The bridge has improved living conditions and has facilitated the placement of equipment inside the bridge.

Sandy or sandy-mud bottom with a water-depth of 50-60 metres are the favoured fishing grounds for this type of boats. Due to the fact that bottom fishes generally prefer to inhabit places where there are abundant demersal organisms (fish food organisms) which exist because of the relationship between topography and sea current, the favourable fishing grounds are in areas such as bottoms with a lot of ups and downs rather than flat bottoms, or areas surrounding reefs and cavities. However, bottoms with many rocks, gravel, seaweeds or shells cannot be used as fishing grounds since net towing will be difficult.

The fishing is carried out in the day-time from dawn to the afternoon throughout the year, as the fishing is meant to catch miscellaneous demersal fishes. But the time period with the highest catching rate is the time before and after tide changes, since, it is the time when the net opening is most stable and fishes start to gather in the area near the bottom. The netting rate is also high around sun-rise when the food searching activities of fishes are more.

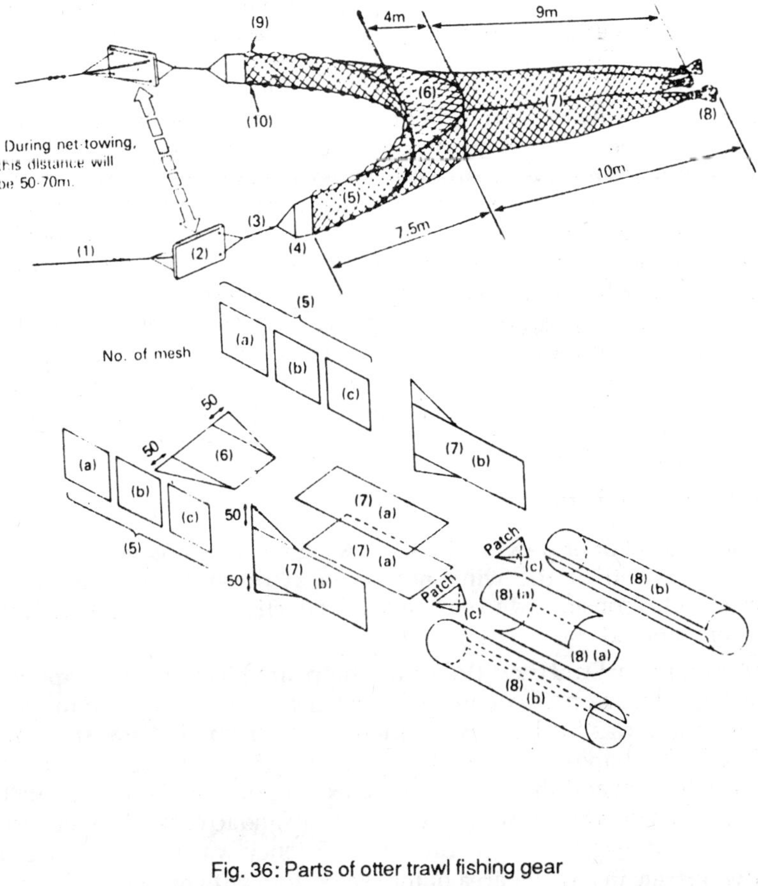

Fig. 36: Parts of otter trawl fishing gear

Basically the fishing gear is composed of the following parts.

(1)	Towing rope	—	Wire rope of 8-9 mm. diameter, 300-400 m. length, 3-4 times of water depth.
(2)	Otter board	—	Made of wood or resin of 125 cm. (length) X 60 cm. (Width).
(3)	Rope	—	Nylon of 35 mm. diameter.
(4)	Supporting pole	—	Iron bar
(5)	Wing net	—	Nylon 12 ply
(6)	Ceiling net	—	Nylon 12 ply
(7)	Belly net	—	Nylon 12 ply
(8)	Bag net	—	Nylon 12 ply
(9)	Float	—	Made of foam plastics
(10)	Sinkers	—	Made of iron or ceramic.

Net casting is to be started at a slow-speed cruise, when net casting proceeds until the end of the wing net is reached, the otter boards are placed in position with chains, and then by extending the warp, the trawling state is created. Trawling speed (speed against water) is 3-4 knots for net catching large fishes and about 2 knots in case of small prawns. In either case trawling is performed from up to down according to the tidal current. Trawling time is 40-50 minutes per operation and 6-7 trawls are performed in a day. Propulsion is stopped for net hauling and the warp as well as the net are rolled up by means of a hydraulic powered net-roller, while the boat is pulled back. The net-roller is a quadruplex type, and the warp is rolled up into the two side-reels. When the otter boards are pulled up to the deck, they are hung on the bitt at the stern after the chain is removed. Then the warp and net are rolled up into the two central reels. The hydraulic roller is stopped when the bag-net has been pulled up to the stern, and then the bag-net is hauled up and placed on the stern deck by hand. In cases, where the catch is large, one should attach a rope to the bag-net part and haul it up using a derrick. The rope tying up the bag end is untied, and the catch is dumped out on the deck. After throwing the debris caught inside the net, the catch is stored in either of the two fish holds on the sides. During the next net casting, while the rudder is placed on auto-pilot during net towing, catches inside the fish holds are sorted.

Handling the Otter Boards

The old type of otter trawl had a short wing net and thus the otter board was attached at the end of the wing net. Even at present, this method is used for shrimps and is considered to be one of the effective fishing methods, called the "out-trigger" method.

In 1925 Vigneron Dahl trawl (V.D. trawl) appeared. This system employs a hand rope which is attached between the wing net and otter board. With the success of this system, there was further developments of otter trawl. A transformation has taken place in the shape of the net. The square part (belly net) became smaller, and in turn the wing net and the cod end were extended. The hand rope functions to create clouds of sands by scraping the sea bottom and to chase the fish into the net centre by intimidating them. The catching efficiency was improved because one could now operate in a wider area of the sea bottom with higher mobility.

The forces which act on the otter board can be divided into two elements. The sheer force functions to widen the net mouth; however, drag force turns itself into the resistance towards the towing force of the boat. It is desirable to design the shape of the otter board in a way so that the sheer force becomes as strong as possible, while the drag force stays as small as possible.

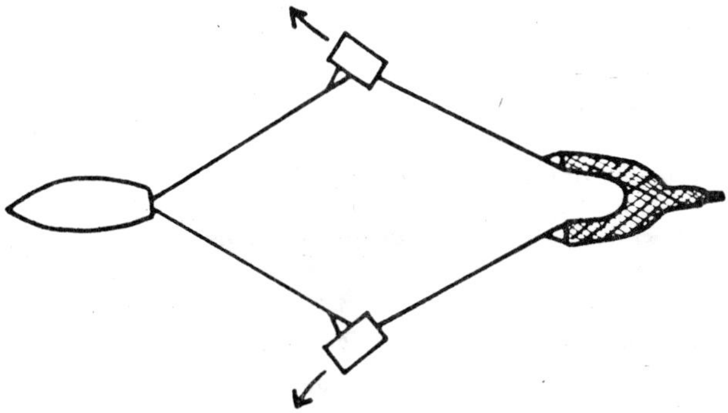

Fig. 37: Forces act on otter board

Although a flat rectangular board is generally used for otter board, several boards of various shapes have been developed to improve its function and put to practical use. Among them are curved rectangular boards, V-shaped boards and oval-shaped boards. The board which are made up of assembled wooden boards strengthened by steel belts are commonly used. However, ones made of duralumin or FRP are also starting to become popular. Measures are taken so that the towing angle can be controlled by devising metal fittings for the pendant chain.

The make-up of the otter board and the attached gear is shown in Fig. 38. Although the board will be moved with A and A′ sides touching the sea bottom, it is necessary to make the chain length adjustable.

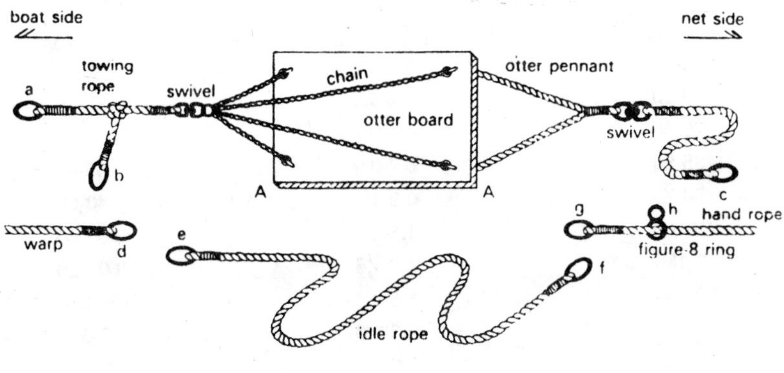

Fig. 38: Otter board with fittings

As regards the otter pennant, the lower part should have several centimetres longer than the upper part, so that during towing, the board will face upward slightly. Swivels should be fitted to both the front and back parts of the board in order to prevent the rope from twisting. The idle rope should have some marginal length by making it several metres longer than the length of the towing rope and pennant rope put together. Regarding the connecting metal fittings, besides the rings with gaps, shackles or G shaped hooks are used where necessary.

Otter Board Specifications

(a) For double rig shrimp trawl boards

Vessel h. p.	*Board dimensions Metres*	*Board weight kg.*
100-150	1.8×0.8	60
150-200	2.0×0.9	90
200-250	2.4×1.0	120
250-300	2.7×1.1	160
300-450	3.0×1.2	220
450-600	3.3×1.3	300

(b) For oval-shaped trawl boards

100-150	2.05×1.21	290
250-350	2.25×1.33	380
350-500	2.36×1.39	460
500-900	2.45×1.4	700
900-1800	3.12×1.78	1000

The oval-shaped boards are slightly longer and deeper than equivalent rectangular boards to compensate for the slight loss of surface area due to the shape.

For rectangular otter boards; ratio of length to breadth 2.2:1

Trawlers h.p.	*Board area Sq. mt.*	*Board Weight Kg.*
10-20	0.4-0.5	12-20
20-35	0.5-0.7	20-30
35-50	0.7-0.9	30-40
50-75	0.9-1.1	40-60
75-100	1.1-1.3	60-80
100-150	1.3-1.6	80-120
150-250	1.6-2.3	120-200
250-400	2.3-3.0	200-350
400-550	3.0-3.7	350-500
600-900	3.8-5.2	500-800
900-1300	5.2-6.1	700-1100
1300-1800	6.1-7.2	1000-1400

Steel Vee Doors L : B = 1.7 : 1

Trawlers h. p.	Board area Sq. mt.	Board Weight Kg.
50-75	1.0-1.2	60-80
75-100	1.2-1.5	80-100
100-150	1.5-2.2	100-150
150-250	2.2-2.9	150-260
250-400	2.9-3.5	260-400
400-550	3.5-4.4	400-600
600-900	4.4-5.5	600-900
900-1300	5.5-6.6	900-1250
1300-1800	6.6-9.0	1250-1600

Beam Trawlers

Beam trawlers are forerunner of otter trawler and most bottom trawlers used beam trawl before the introduction of otter trawl. Except for some shrimp, flatfish and sole fisheries, beam trawling is not used much today.

Beam trawls are cumbersome to use and are almost always fished over the side. The old sail boat trawlers of the North Sea used a heavy beam trawl of 20-35 feet width (6-11 metres). It was set over the side while the vessel sailed before the wind, but with it more on the starboard quarter. A heavy towing post held the warp when towing and a capstan was used to winch the gear in.

Modern beams are made of steel. Several tickler chains may be added near the foot rope to increase the weight of the gear. It is common practice to work with two beam trawls from booms on each side, similar to that of double-rig on shrimp trawler, and the booms must be strong and well stayed to take the loads. Safety release systems are sometimes incorporated in the boom stays or winch brakes to prevent the boat from capsizing if the gear snag suddenly on the sea bed.

Small beam trawl, with various improvements, concerning the number and arrangement of nets towing by one fishing boat and construction of nets are in operation in Seto Island Sea, Japan to catch shrimp. There are two methods of net towing, one net towed by one fishing boat and two nets towed by one fishing boat. There are three kinds of nets; a net with one bag, one with two bags and one with three bags.

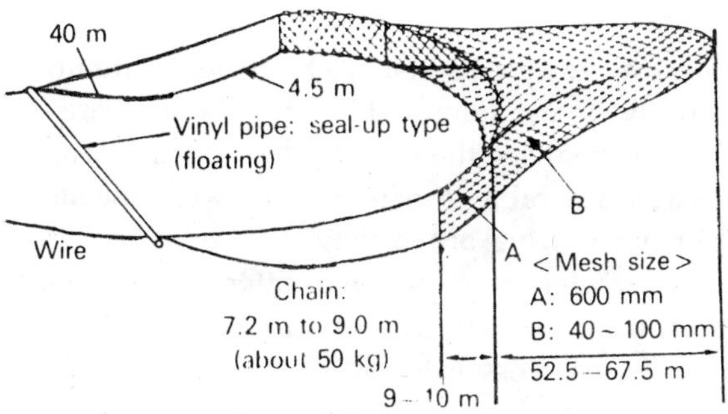

Fig. 39: Beam trawl

Construction of beam trawl

Warp	– 200-300 m.	Cod end	– 5 m. for shrimp
Bridle	– 50 m. each	Cod end	– 2 m. for debris.
Wing	– 90-270 cm. height, 60-70 m. length.	Beam	– 10 m. length
		Float	– 16-17 floats made of styrofoam.
Bag	– 10 m. length.		

The trawl has an elaborate construction. The codend is divided into upper and lower portions, the lower portion for collecting debris; while the upper portion for shrimps.

Fishing gear (small beam trawl) is laid piled up on the quarter deck after assembling all parts of the trawl except the beam. At the fishing grounds, while running the boat at slow speed along the direction of tidal current, the bag net and the wing net are thrown into the water, when the beam is attached to the base of the bridle with a hook. After the beam and bridle are thrown in, the boat is moved forward at full speed while letting out the warp, till it stretched and the behaviour of the net adjusted. Subsequently the boat's speed is reduced and the net is towed along the tidal current at a slow speed of 1-2 knots for 45-60 min.

Because the shrimps move out of sand for seekng food during night, fishing generally start at about sunset. During towing, the lelt and right wings scrape the bottom, which scares the shrimps and drives them out to the centre of the net and finally scooped up with the bag net.

During net hauling the warp is wound around a side roller driven by engine directly with the aid of a roller at the stern. After the bridle appears, the net is hauled in with side rollers on both sides. The beam is then removed and the net is hauled in by hand. When the net is too heavy to haul in by hand, winch is used.

Fishing operation is done by two persons generally, but with remote control system, fishing can be done by one person also.

Pair Trawlers

Pair trawlers are towed by two boats. The boats maintain a distance apart equal to about half the warp length. No otter boards are needed for pair trawling operation. This reduces the overall drag on the gear and permits the use of larger net. Long heavy cables are inserted between the net and warp end at which point a weight and a swivel are usually attached. There are almost as many different types of pair trawlers as there are of otter trawlers. Pair trawling fleets currently operate in Europe, Scandinavia, North America, Japan, Taiwan, Korea and Thailand. Some of these vessels work for shrimp, some for hake, some for herring, some for cod, some for flatfish and some for mixed species. A pair trawler may be an open boat with an outboard engine, or any vessel up to a 60 metre, 2000 h. p. stern fishing vessel.

Fig. 40: Pair trawlers

For most pair trawlers, a stern trawler or side trawler deck arrangement is appropriate. The larger pair trawlers fish over the stern and the smaller ones mostly over the side. A Kelly's eye and keep arrangement may be used at the warp end weight. Many pair fishers pass the bridle or cable ends from the assisting vessel to the hauling vessel at that point and leave the weight hanging in the gallows from the end of the warp.

Differences in operation arise with Japanese Bull trawlers and Canadian Pair seiners both of which use cables of relatively large size or thickness. Special reels or large winch drums are needed to wind on these cables. Bull trawlers usually pass both the warp ends to the hauling vessel which pulls in all the gear, Canadian pair seiners each haul in all of their warp and cable and pass only the net wing to the hauling vessel. Most other pair trawlers pass the bridles or cables to the boat that set and will haul the net. Apart from those two types, most pair trawlers differ little from otter trawlers.

The pair trawl is set out from the side or the stern of the boat as on an otter trawler. Once the net is in the water and safely clear of the setting vessel, the partner boat comes alongside and throws or receives a light heaving line. The ends of the bridles are tied to the heaving line and are pulled quickly aboard the partner vessel and connected to its warp. Bottom pair trawlers mostly use a single warp each, while midwater pair trawlers mostly use two warps from each vessel.

Both boats take the weight of the gear and the skippers check to see that the net is not fouled in any way. Then at a given signal the winch brakes are released and the bóat shoots their warps in separate directions. Bottom trawlers usually set the gear out in 'U' shape, while mid-water trawlers set theirs in a 'V' shape. Because in bottom fishing pair trawlers like to sweep as wide an area of the sea bed as possible. When all the warp is set and the winch brakes applied, the trawlers commence towing on parallel courses.

It is important that the skippers keep in close touch with each other during the operation with the help of vhf or walkie talkie radio.

The hauling operation varies with the type of pair trawl gear in use. Most pairs heave in all the warp, drawing gradually closer as they do. When both warp ends are up, a heaving line is thrown across and the bridle ends from the assisting vessel are attached to it. The bridles are released and the hauling vessel pulls them in and proceeds to haul both sets of bridles together. The release of the bridles from the warp end is best accomplished by means of a quick release hook.

To avoid danger of collision due to the close proximity to each other, both skippers must be alert and must have direct control of the steering gear and engine at all times. For both boats to maintain proper steerage, the warps are always towed on the starboard side regardless of the side on which the gear or other vessel lies. Small boats may tow from a central point on the stern.

If one vessel has more power than the other then the stronger vessel will take the starboard side of the gear as it requires more helm.

Danish Seiners

There are three basic type of bottom seiner; Japanese seiners, Danish seiners and Scottish seiners. Danish seiner is another type of bottom trawler and do not uses otter boards. Instead it sets over a mile of rope in an arc or 'L' shape on each side of the net. The net itself resembles a large light trawl with long wings. The rope warps scrape slowly over the seabed, all the while, driving fish in towards the advancing net. Once the ropes are almost closed or parallel to each other, the vessel speeds up to drive the remaining fish into the net and then the warps are winched quickly in. The whole operation takes from one to two and a half hours and in that time an area of the seabed measuring over one mile by half a mile will have been "swept" clean of fish consisting cod, had dock, hake, pollock, flatfish, skate sole and prawns.

While Danish seiners use an anchor and only winch in the gear, Japanese seiners tow the gear like a trawler and Scottish seiners tow and haul in the warps (very slowly) at the same time.

The main items of deck equipment are the winch, which resembles a large 2 drum capstan, and the reels on to which the long bulky rope warps are wound.

Danish seiners vary in size from 16 to 30 metres in length, and from 120 to 650 h.p. main engine. They are most effective in temperate waters, where there is abundant demersal fishes.

Trawl Nets

There are hundreds of variable trawl net designs to suit each particular class of vessel, but basically, dealing with the deep sea type of gear, this can usually be traced back to resemble the original Granton Trawl net pattern. But since past fifteen years a good deal of original thought has been put in designing trawl net to suit different size of stern trawlers.

Otter trawl nets were developed from the beam trawl and designed with seven main pieces.

The Square

The section of netting fitted between the top body and the two top wings so that it partially overhangs the groundrope. This part of the trawl net is most important for it does to a great degree govern the way the rest of the net can be shaped. The specification of the piece is 260 meshes of double twine are set up in a $5^1/_2$" to 6" mesh and two double rows are braided. Continuing with single twine, baitings are made at each side every fourth row decreasing to 200 meshes across. If a $5^1/_2$" mesh is required, the finished length would be approximately 28 feet, whereas a 6" mesh would result in a length of 30.5 feet. The cutting rate on machined netting would be 1 point 2 bars repeat.

The Top Wings

Two sections of netting usually shaped diagonally opposite to one another to form the upper mouth of the trawl. The headline is attached from one top wing end along the diagonal flymeshes, across the bosom or centre part of the square and along the opposite top wing to the end.

The quarters are the two areas where each of the top wing flymesh edges adjoin the square. With conventional trawl nets the quarters are usually subjected to heavy loading due to inadequate shaping, and this is particularly true for such a trawl net as is operated from a side trawler when the gear often has to be "shot away" with uneven strain on the headline.

Specification: Ninety meshes are set up in $5^1/_2$" or 6" mesh making 15 or 20 in double twine at the inner flymesh side which forms the double twine at the inner flymesh side. This forms the "double corner" (quarter area) which is usually made first, then the wing is completed to 11 meshes wide mostly in a single twine following with flymeshes from the double corner. This reduces the width by one mesh for every two rows braided, as the outer selvedge is plain, which means, if there is no baitings in the "quarters" then the finished length would be 36.5 feet in a $5^1/_2$" mesh or 40 feet in a 6" mesh. However it is normal practice to make sixteen or so baitings in the flymesh edge, thus the length would be reduced to 29 feet and 31.5 feet respectively. It is also the practice to include in the finished length several double rows at the wing end. And occasionally two double meshes may be braided along the entire flymesh edge. In each case the purpose is to gain added strength for the more vulnerable parts of the netting.

Lower Wings

Two narrow sections of netting fitted between the lower belly and the top wings to form the lower lip of the trawl net. The ground rope is attached from one wing end to the other, along the flymesh edges and across the lower belly bosom meshes. The two lower wings are the parts of the netting which are subjected to the most abrasion, and consequently there are the sections which have to be continually repaired or replaced when working on rough ground.

Specification: Fifty meshes are set up in a $5^1/_2$" or 6" mesh making 10 in double twine at the inner flymesh side for the double corner. Flymeshes are made and the outer selvedge is creased every third row, thus in $5^1/_2$" mesh to reduce from 50 to 17 meshes, which is the required width, the length would be approximately 45 feet as the loosing rate is one mesh for every six rows made.

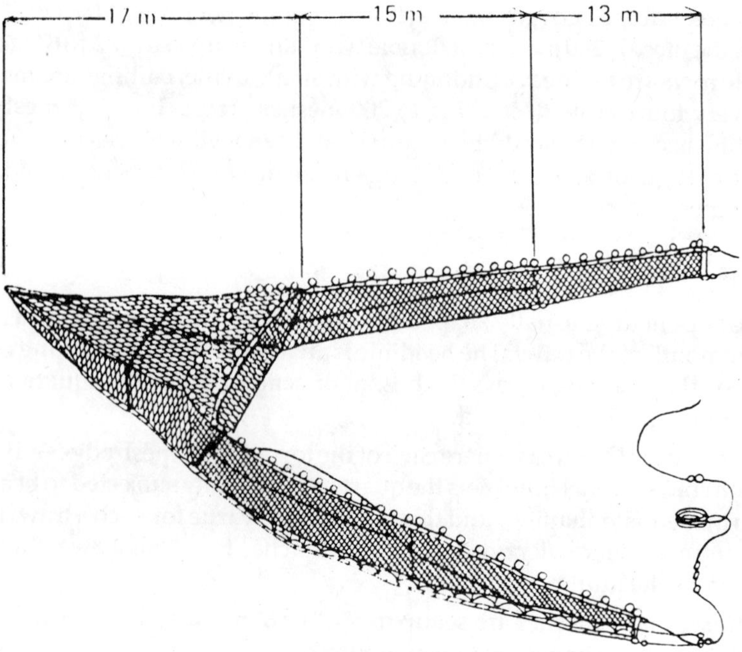

Fig. 41: Parts of trawl net

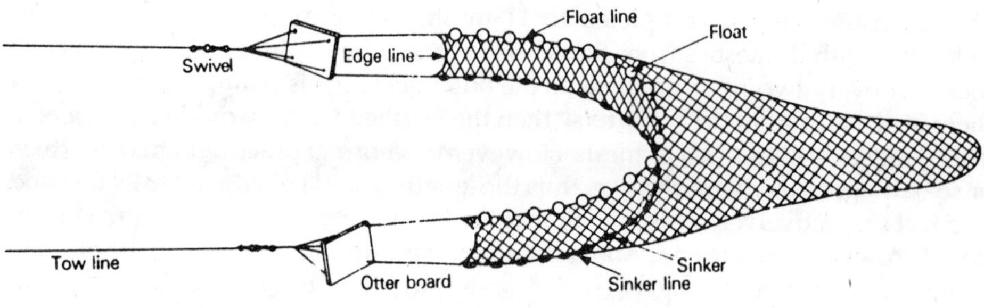

Fig. 42: Assembly of trawl net fishing gear

In a 6" mesh the length would be 49.5 feet but, irrespective, a double row has to be made at this juncture to mark the part where the wing end and section begins with 17 meshes maintained throughout by fly meshing and flat baiting (baiting every two rows) to the required length which should be 10 per cent greater than the total stretched length of the square and to wings, including several double rows at the wing end. Again, double meshes may be made along the entire flymesh edge as added strengthening.

The "bunt" is the netting between the bosom corner up to the end of the bunt bobbins or bunt section of the ground rope. In other words a double row should be made or a coloured mark inserted for several meshes at a position approximately 15 per cent greater on the netting than the length of the bunt bobbin wire or similar part of the assembly. In case of a 20 feet ground rope or footrope wire being in this position the marker would have to be 23 feet from the start of the lower wing which would leave about 5 per cent slack netting to be spaced along the remainder of the ground rope.

The Bellies

The bellies are the channel of the trawl which guide the fish along into the codend. Two are required for each trawl and the top one used to be referred to as the "baiting", whilst the lower was regarded as the belly of the trawl. Conservation aims and changing minimum mesh legislation have meant that this part of the trawl net has continually had to be changed. This compulsion has not, in fact, always been as bad as it sounds for the increases demanded have led to the trawl net being gradually enlarged. This has happened over a number of years and at the same time the size, range and power of vessels has steadily increased.

Depending on the twines used, the minimum mesh size (inside measurement) is between 110 and 130 mm. for the normal deep sea regions, which the heavy British trawlers work. Assuming that a $5^1/_2$" mesh is worked in the mouth of the trawl net, this should be continued along the entire length of the bellies, whereas a few years ago the belly was graduated into a number of spools of decreasing mesh size. For practical purposes, however, a pair of bellies can be considered to have an overall length of 50 feet with approximately a $5^1/_2$" mesh throughout.

Specification: Two hundred meshes are set up with double twine and 2 to 10 double rows, or in certain cases even more, are braided, and continuing with single baitings are normally put in every third row at each side down to say 50 meshes. The length therefore, of each belly would be, with a $5^1/_2$" mesh, 51.5 feet (225 rows, 15 baitings every third row). Although there are no changes of "spools" it is advisable to have double rounds or several coloured marker meshes dividing the belly up into sections in order to facilitate repairs.

Lengthening Pieces

Originally, lengthening pieces simply served the purpose which the name implies. They were 10 to 20 feet sections used to extend the bellies as required. Now-a–days however, this extension may be allowed for in the belly length or it can be used to advantage in certain other cases.

Specification: Using double netting for strength an extension of about 15 feet might be added to make a section from 66 to 50 meshes decreasing with baitings made every eighth row at each side. Total number of rows 64.

Cod End

Now-a-days cod ends are two rectangular pieces of netting manufactured with heavy double twine. The top edges are joined to the narrow end of the bellies, the selvedges are laced together and a codline or codend clip is reaved through the lower meshes for securing the section into a bag where the fish are held until released on board the trawler. It is the codend which are most affected by mesh legislation and the size of the codend depends on the size and type of vessel, the area to be fished and the species to be caught.

Specification: The minimum inside mesh size allowed to be worked in the North Atlantic deep sea fisheries varies between 110 and 130 mm. Also a stern trawler will work a longer codend than a side trawler. For practical purposes, set up 50 meshes of double twine and braid 174 rows with $5^1/_2$" mesh plain down to 50 meshes.

Flapper

The flapper is a small single piece of netting tapered and fitted into the bellies near to the codend so that after the bellies have been sealed with twine diagonally towards the selvedges with the top edge of the flapper laced to the top belly it acts as a non-return valve after fish have entered the cod end.

Dredge Net

A dredge net has a wood or steel frame with claws made of steel or other metals that scrape the seafloor as the net is towed. Usually 2 to 5 of these nets are used together to take bottom fish, shell fish, sea slug etc.

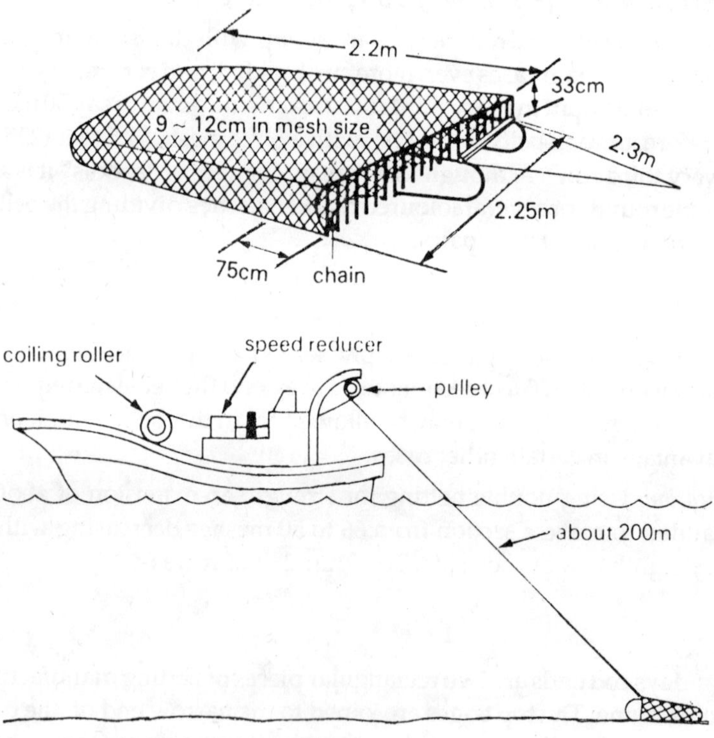

Fig. 43: Dredge net

In many ways the dredge net resemble beam trawlers. The net are small steel beam trawls with metal rings instead of nettings on the underside. They scrape fish and shellfish from the sea bed. Large vessel may work three dredge nets on each side, from booms as on double-rigged beam trawlers.

The fishing gear is divided into several types which differ from one another in operation method, weight, shape and size of opening, and mesh size of a bag net. Three types of dredge net are currently operated in the sea.

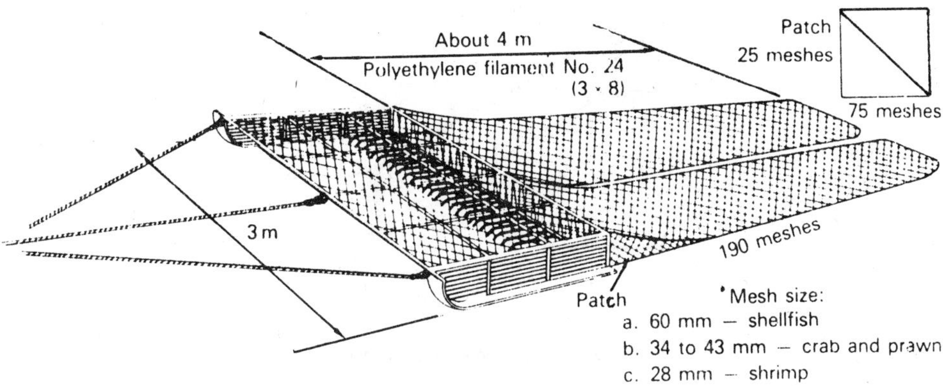

Fig. 44: Dredge net with iron teeth edge

An iron frame which has a number of iron teeth on its lower edge is fitted to the opening of a bag net.This assembly slides over the sea floor by means of a pair of runners fitted to the fore part. Wire netting over these runners prevents the assembly from sinking in muddy sand while sliding over the sea floor. This fishing gear is used to collect a variety of fish, crustaceans and shellfish. The mesh size depends on the catch sought.

Fishermen reach a fishing ground within 30 to 45 minutes after leaving the port early morning. A sea floor of muddy sand at a depth of 17 to 20 metres is suitable for the operation of this type of gear. Towing operation is continued for 30 min., while the boat is moving at a speed of 2 or 3 knots. Within 10 to 15 minutes net casting and

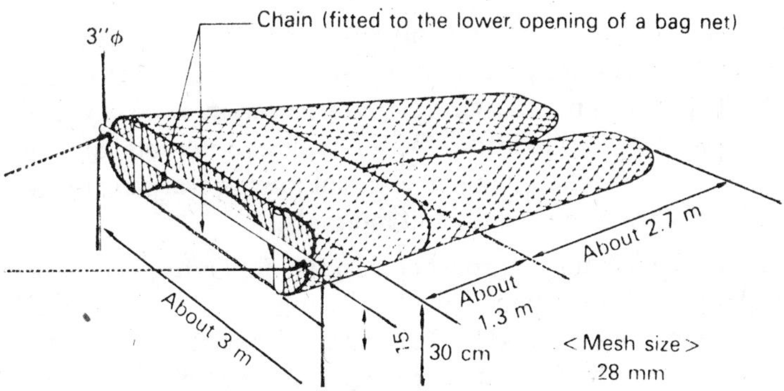

Fig. 45: Dredge net with chain

hauling operations are completed by means of a power-driven device. Duration of daily operation is usually 6 hours, during which towing operation is repeated 12 or 13 times.

In the second type, the iron frame has a special chain device in place of the iron teeth. This device scrapes the sea floor to dig out fish and crustaceans, especially crabs.

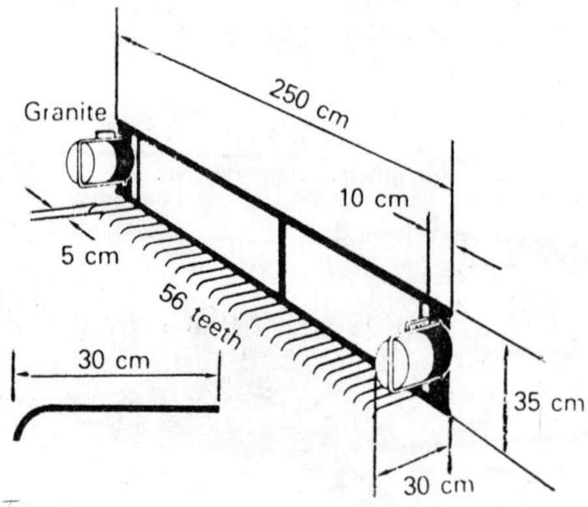

Fig. 46: Dredge net with iron teeth on the lower edge

In the third type has a number of iron teeth on the lower edge of the frame. But it features special weight in place of the runners. Two bag nets are fitted to the frame by means of a pole. These bags jump alternately through the balancing action of this pole, while they are towed along the sea floor. Net design is the same as that of first type and mesh size is 60 mm.

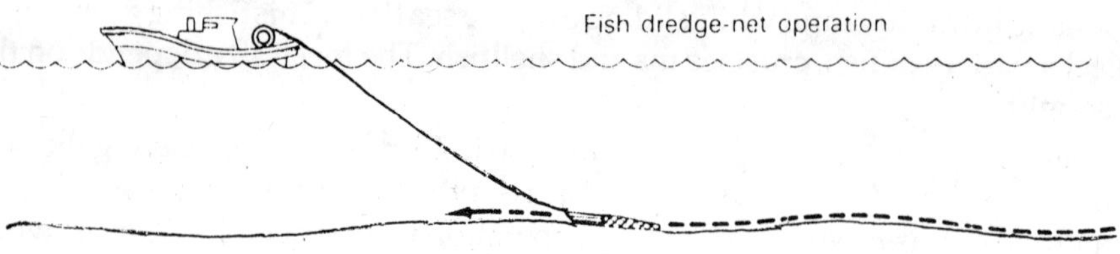

Fig. 47: Fish dredge net

If the desired catch is demersal fish or crustaceans, the iron frame with dredge teeth slides over the sea floor by means of a pair of runners fitted to the forepart. In the case of shellfish collecting, special weights are used in place of runners, so that the iron frame assembly is lifted from the sea floor at intervals (Fig. 48). Several different means including changing the length of towing warp are in use to lift the iron frame assembly from the sea floor at intervals.

The weight of the fishing gear must be decided depending on the horsepower of engine and the towing capacity of the fishing boat. Dredge net fishery in the inland sea of Japan is operated by 3 to 5 tonne class fishing boats driven by up to 15 h.p. engines. The dredge net fishing gear is available in several different weights: 100 kg., 130 to 150 kg. , 200 kg., 400 kg. and 500 kg.

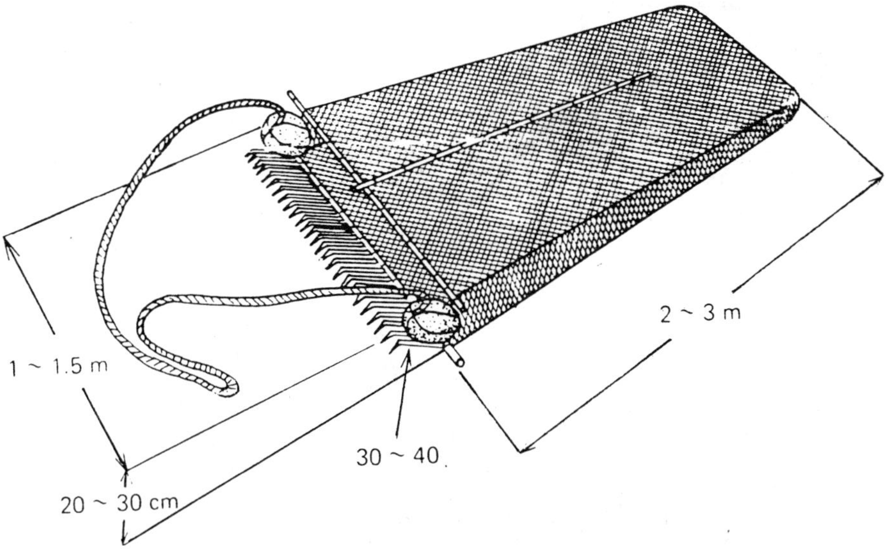

Fig. 48: Dredge net for demersal fish and crustaceans

In most cases the overall width of the iron frame assembly is 250 to 300 cm. The overall height of this assembly is 20 to 30 cm, when fitted to a fish or prawn dredge net, and 30 to 40 cm. when fitted to a shell fish dredge net, because in the case of later the bottom of the assembly sinks deeper into the muddy sand than in case of the former.

The bag net presently in use is made of polyethylene material. The mesh size varies depending on the catch sought,

Shellfish species – 60 mm.

Demersal fish species – 35 to 43 mm.

Prawn species – 28 mm.

The net is cast and pulled up from the stern of the fishng boat. The towing warp is rolled in by (a) a line hauler (driven electrically or hydraulically) mounted on the stern deck or (b) a side drum connected directly to the main engine.

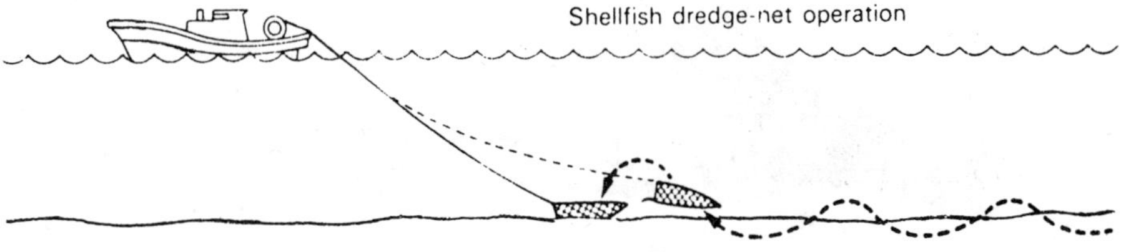

Fig. 49: Shellfish dredge net

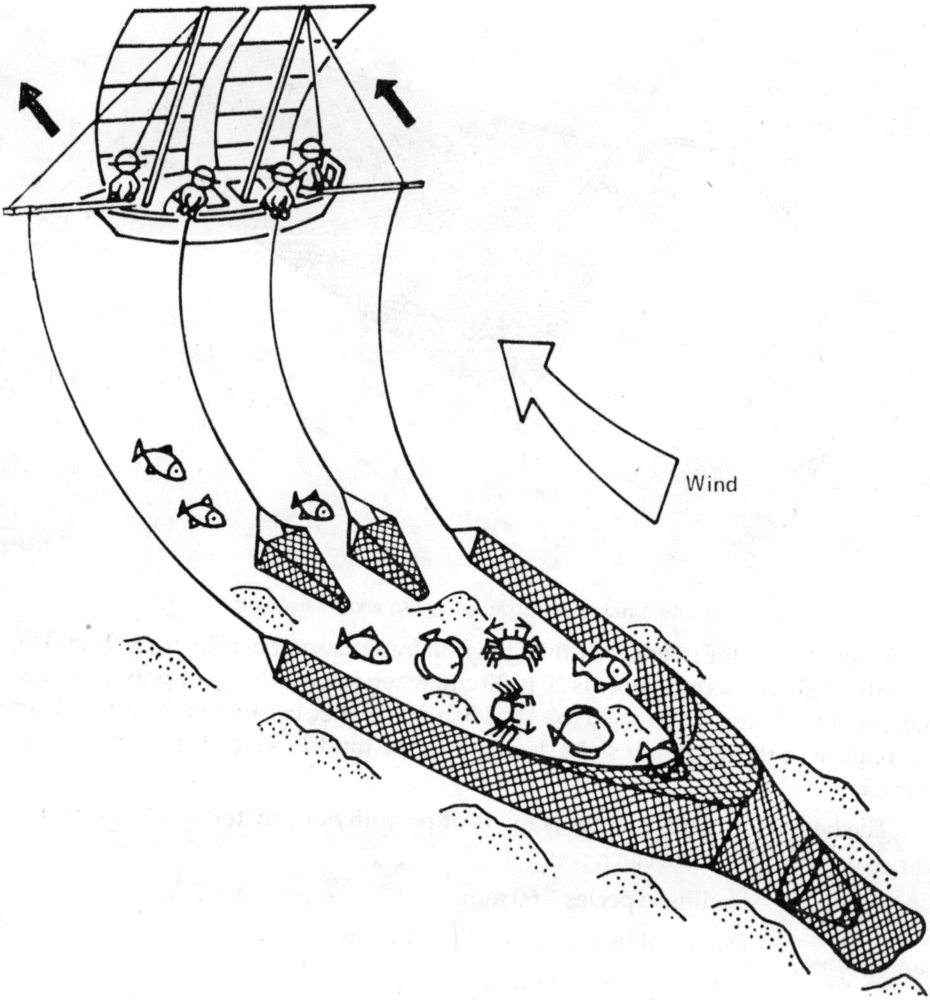

Fig. 50: Sail trawling

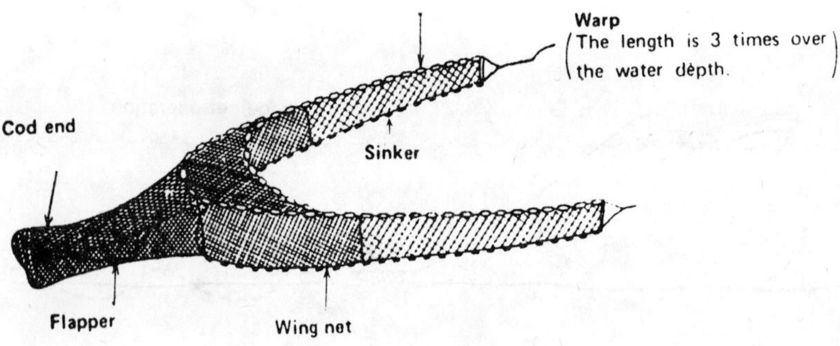

Fig. 51: Trawl fishing gear

The towing warp tied to the side roller on the starboard tows the dredge net along the sea floor via a cylindrical spring device which is hung down on the side of the bridge. This device gives an intermittent impetus to the towing warp, which in turn lifts the iron frame assembly from the sea floor.

Sail trawl

This is a drag net; but the net is pulled while the fishing boat is slipped sideways using manpower, wind power or current, an unique method. Long poles are attached to the front and rear of the vessel to increase the range. To these are attached 1 to 5 small nets of various sizes and patterns.

In the shallow seas, this method is used especially for crab, shrimp flounder and flatfish.

5

BOAT SEINE

Boat Seine

Boat seining involves (a) a wing net or ground rope, (b) a main net, (c) a bag net and (d) a tow rope. These parts have the following functions:

(a)	Wing net (ground rope)	–	This net is used to intimidate the fish so that they gather into a school.
(b)	Main net	–	This net prevents the fish from escaping and guiding them to the bag net.
(c)	Bag net	–	This net holds the fish.
(d)	Tow rope	–	This rope is used to handle and tow in the net.

The shape of the nets and gear used in boat seining vary a great deal and these differences in the size and shape of the nets used are a result of the habits and characteristics of the particular fish that is being caught. When there is a change in

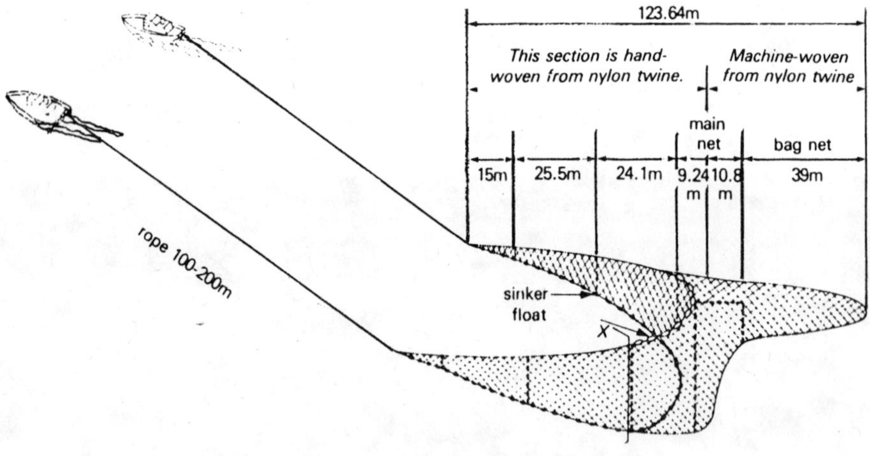

Fig. 52: Pair-Boat seining

the type of fish to be caught, either certain parts of the net may be changed to suit the habits and ecological characteristics of that fish or else the whole scale of the fishing gear may be changed along with an accompanying change in the boats motor capacity.

A given fishing method involves the unification of three basic elements, the target fish, the fishing gear, and the fishing boat. The shape and size of the nets used in boat seining varies a great deal depending on the particular conditions of the fishing grounds, while at the same time, sharing the same structural and functional characteristics.

Anchor Seining

Anchor seining is performed with small nets using the tidal current force, and requires much less fuel than trawling.

The fishing ground is a flat sea bottom with a strong tidal current (2-3 knots). The most suitable bottom composition is one of small gravel. Sandy or muddy bottoms are unsuitable because the net can get stuck in the bottom, thus hampering the towing. The appropriate water depth is 10-40 metres.

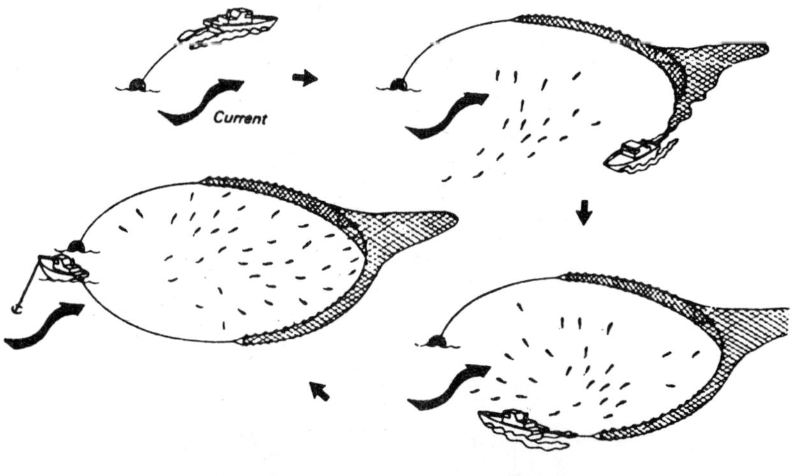

Fig. 53: Operation of anchor seine

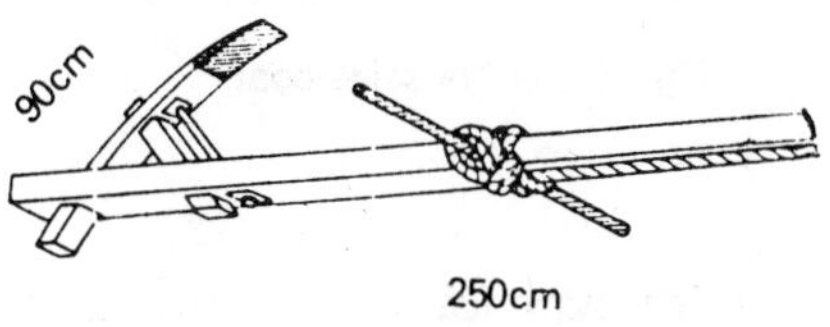

Fig. 54: Anchor

The catch consists of octopus, shrimps/prawns, mantis shrimp and demersal fishes such as flatfish, rock trout and small sea bream. The fishing boat has an overall length of about 30 feet with diesel engine of 15 h.p.

First the warp with a float attached to the end is cast into the water, then the warp, ground rope and the net are strung out in succession with the boat moving in a circular motion at full speed. After running a circle of about 200 metres diameter, the boat returns to the starting point and picks up the float. The boat continues to tow for 20 or 30 seconds to adjust the stretch of the net. When tension comes to bear on the warp the boat is stopped and the anchor is cast. Next the net hauling begins. A single net operation takes from 25 to 30 minutes.

The net hauling is performed by a standing-type roller made of wood (pine), that is turned by power from the boat engine. The roller is made with 2 layers so that two ropes can be hauled simultaneously but separately. During hauling the boat's engine is put in neutral and the roller is turned at the lowest speed (400-450 r.p.m). After hauling up the net until the wing nets can be seen, the roller is stopped and the net is pulled up into the boat by hand. This operation is performed by two people.

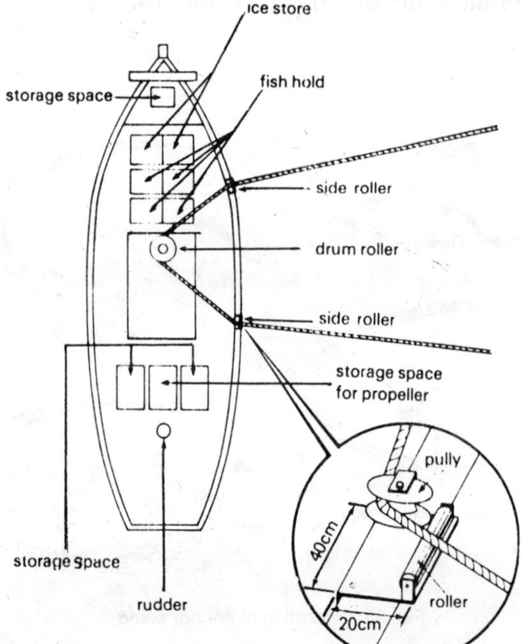

Fig. 55: Anchor seine boat

Pair Boat Seining

Two fishing boats of identical type leave for fishing grounds with their bows linked. On arriving the fishing ground, a fish finder is used to locate a high density school of shrimp upon which the net is cast.

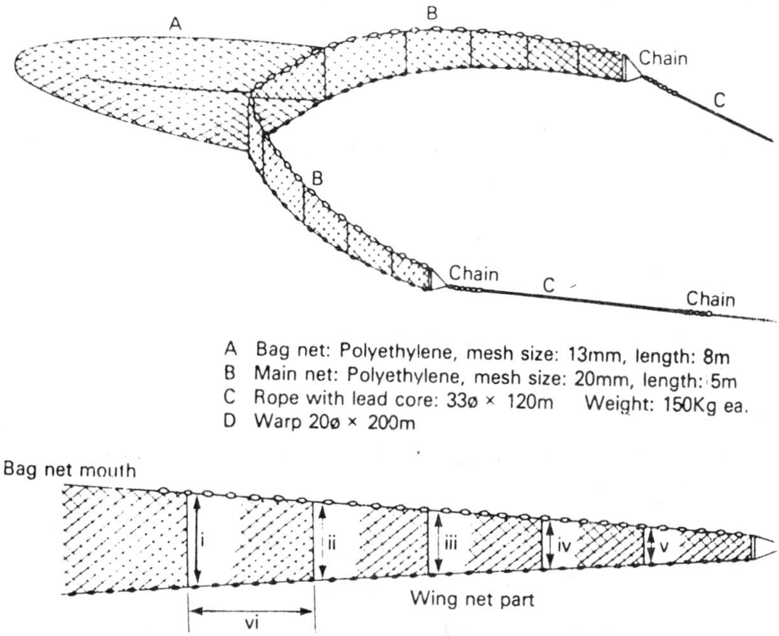

Fig. 56: Seine net in anchor seining

The rope linking the bows is untied and then the bag net and main net are cast as the two boats move away from each other at full speed. At this point the net zonde transmitter is attached to the ground rope and dropped into the water. Thereafter the wing net and tow rope are cast.

During towing the tow rope should be attached to the wire of the tow rope fulcrum controller so that the direction can be changed when necessary. The length of the tow rope is generally 1.5 times the depth of the tow net. The distance between the boats should be kept some 100 metres. The towing is done from up to down according to the tidal current, and the towing speed is about one knot.

The depth of the towing net and the altitude of the net mouth are the critical factors. While observing changes in the depth of the shrimp concentration with a fish finder, the combination of the towing speed, the tow rope length and the distance between the boats is adjusted as necessary. The depth of the ground rope and the net can be constantly monitored by means of the net zonde.

The duration of a single net towing is approximately one hour. With radio contact the two boats approach each other and after their bows are linked, the net hauling start. During net hauling the boats continue to move forward slowly. The tow rope and the wing net are rolled up on to the net roller. During this operation, although the bows of the two boats are linked, there is a distance of several metres separating the two sterns, so the bag net is drawn up into the water surface between the two boats. While sinking away from the water surface, the bag net is drawn in gradually by hand and when the catch appears at the surface, it is scooped up with a large spoon net, which is hauled up with a winch and landed on the deck. This operation requires 8 persons on each boat. The net is usually towed 5 to 6 times during one night.

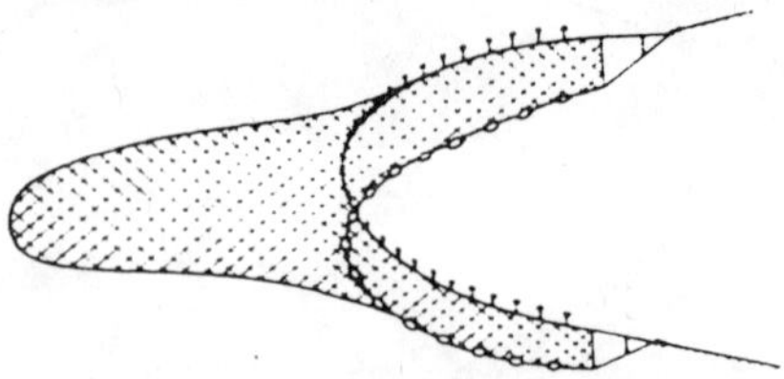

Fig. 57: Seine net

Boat Seine in Japan

The fisheries concentrate mainly on catching mature anchovy and sardine of fingerling to half-grown size.

Sardine lay eggs in the months of December to February. The water temperature in the spawning areas ranges from 12°C to 19°C. In the warmer sea areas and the coastal waters the laying of eggs occur early from December to February, while the remaining majority of eggs are laid from February to May. Eggs are laid in areas where warm currents mix with the coastal waters. After hatching the fry gradually gather into schools and live in coastal waters with an abundance of the planktom which the fry feed on. During the fry and fingerling stages of their life cycle, the sardine live at a depth of 20 to 30 metres in the coastal waters having the optimum water temperature. As they grow to fingerlings, they begin to ascend branches of the Kuroshio current into the Inland sea from both the channels that enter from the east and west. During this phase boat seine fishery is conducted in the open sea and in the channels to the Inland sea to catch these fry. The sardine continue to grow rapidly from spring to early autumn, and the schools of half-grown to mature sardines form fishing grounds in the Inland sea.

In autumn the sardine yearlings avoid the drop in water temperature in the Inland sea by migrating once again to the open sea.

Anchovy live in areas with the same type of water conditions as the sardine. Though the life cycle of the anchovy follow the same general pattern as that of sardine, yet anchovy are able to live in areas with larger range of water temperatures, from 10°C to 30°C, and whereas, the sardine will migrate over large areas of ocean, the anchovy will stay in one area migrating mostly from deep water to shallow water and back again, making their life pattern different from the sardine.

Anchovy born in the open sea spawning grounds during March and April (spring population) migrate to the coastal waters 2 or 3 months later than the sardine and proceed to migrate into the Inland sea in much the same way as the sardine, when they mature. In the case of anchovy some mature schools will spawn in the Inland sea during the summer or autumn season, resulting in the addition of a summer born population or an autumn born population to the fishing rosources. In other words, populations born in the different seasons and migrating in different areas form fishing grounds in all the areas almost year-round.

The bows of two fishing boats are tied together and the two share the load of the net as they head for the fishing ground. After reaching the fishing ground and locating a school of fish by means of a fish finder the boats carrying the net position themselves up-current from the school. Then advancing with their bows still connected the two boats first lay out the bouys and then the bag net into the water. At this point the rope connecting the two bows is released and the two boats head off in opposite directions to the right and left at full speed, laying out the main net and the wing net as they go. When all of the nets are in the water the two boats turn 90° and begin to run parallel to each other in the direction of the school laying out the tow rope until it goes to the right length that will bring the mouth of the net to the same water depth as the school of fish.

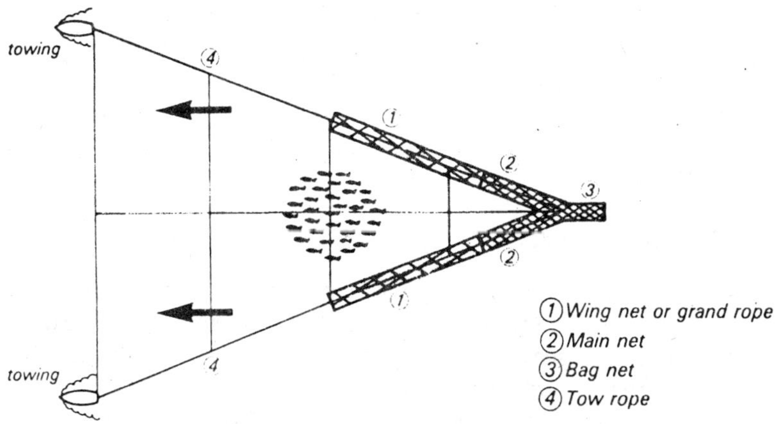

Fig. 58: Boat seine from large motorized vessel

When the proper depth is reached the towing begins. To begin the net hauling operation, first of all the two net boats come together and their bows are once again tied together, after which the net haulers that are positioned on each of the boats are used to reel in the tow rope and wing net on to the reels. To pull in the main net the crew stand on the sterns of the boats and haul it in by hand. When the bag net reaches the surface a transport boat is called up to the stern of the net boats where it picks up the bouy and hauls in the bouy rope to pull up the tail of the bag net. The crew of the transport boat untie the tail of the bag net and dip the catch out from thereby means of a dip net and loaded into the hatch of the transport boat. The transport boat takes the catch from each net hauling one by one and back to the port to unload.

In case of adult anchovy and sardine fingerlings, the netting of the fish school should be done as quickly as possible. During towing, shiftng in a lateral direction once is allowed, but towing should not be done in a zig-zag manner. Adult fish have a strong schooling instinct and they tend to swim along the wing net and gather in

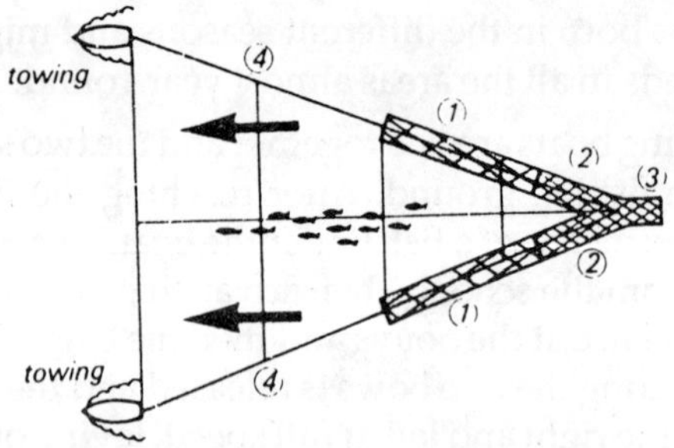

Fig. 59: Small boat seine

the bag net quickly. The towing time depends on the schooling conditions, but generally 10 to 20 minutes or at most 20 to 30 minutes per towing. The towing speed should be 1.5 knots and 7 to 8 towings can be done in a day.

Since the anchovy fry and sand lance tend to scatter, the school should be chased by towing the net in a zig-zag manner. One towing will last for two hours and the towing speed should be 2 knots with four towings in a day.

As the mackerel and horse mackerel have a tendency to escape by swimming in a forward direction after entering the net, this type of net is not suitable for catching those fishes.

Boat Seine for Demersal Fishery

The boat seine with high ratio of shrinkage in the central section of the net cause the net expands into a large bag shape, being pulled by the tide. This type of net is towed along the reefs or near sea bottom to catch the demersal fishes living in rough or rocky bottoms where it would be impossible to use a trawl net. This is a method with a high selectivity of catch, and it includes a number of variations depending on the type of fish being caught.

The seining can be performed either by one or two boats, and the operations can be divided into four basic variations:

(1) Hand hauled net operating from one boat (5 tonne or less/18 h.p. engine or less).

(2) Roller-hauled net operating from one boat (5 tonnes or less/18 h.p. engine or less).

(3) Net operated from two boats (5-15 tonnes/18-35 h.p.).

(4) Net operated from two boats in off shore waters (5-15 tonnes/35-70 h.p. with fish finder).

There are two basic types of net, one being a flat net (without bag net), which billows into a bag like shape as it is pulled, and the other which actually has a cylindrical bag net attached to the central part of the main net. However, the basic operation of both types of nets is the same.

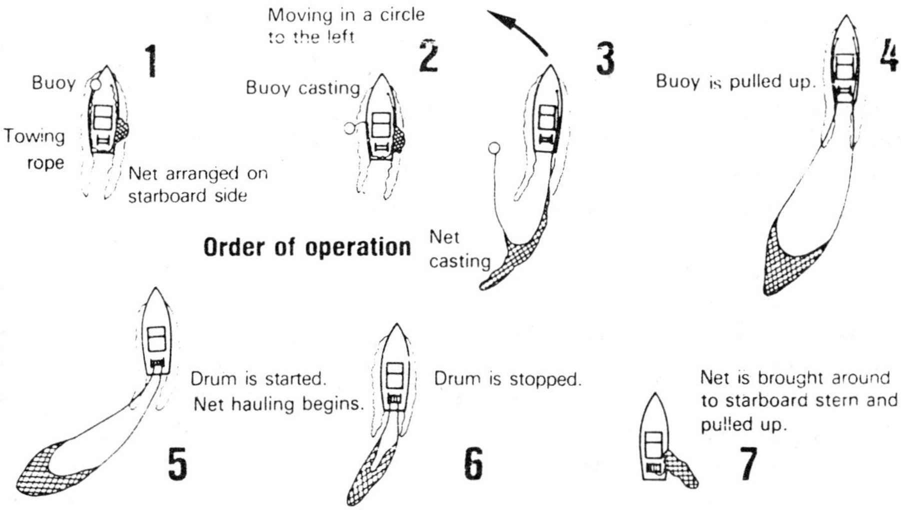

Fig. 60: Two-boat seine operation method

The nets used by two boat operations are roughly twice the size of those used in one boat operation. The catch effectiveness of the net operation, namely, (a) the sectional area of the mouth of net (the width and height of the mouth of the net during towing), (b) the area over which the tow is made (the total area encompassed by the net during towing), (c) the backward speed of the boat during net hauling, (d) the speed at which the net is hauled in, and (e) sweeping distance (the distance that the net moves while it is being closed) are the factors which correlated most directly to the number of fish caught by one towing, were the sectional area of the net mouth and the area over which the net was towed.

The various type of boat seine use different areas of the fishing ground according to the following patterns:

(i) Hand-hauled net operated by one boat in coastal and bay waters with a depth of 20 metres or less.

(ii) Roller-hauled net operated by one boat in coastal waters.

(iii) Net operated by two boats in sea areas with a depth of 40 to 100 metres.

(iv) Off-shore two boat net operation in waters with a depth of 100 to 120 metres.

The following types of fish are caught by the boat seine:

(a) Silver whiting operating from one boat.

(b) Sea bream family operating from one boat.

(c) Lizard fish family operating from one boat.

(d) Threeline grunt family operating from two boats.

(e) Red sea bream family operating from two boats.

Originally one boat hand-hauled method was developed for catching primarily silver whiting in the shallow sea areas. A few red sea bream fingerlings were caught along with the silver whitings. With the demand of red sea bream fingerlings for seeding fish farms, this boat seine operation specialized in catching red sea bream fingerlings alive. Its catch includes sea bream of all ages, from fingerlings to spawing adults (fingerlings of 5cm. to six or seven year old fish of 60 to 70 cm.)

The red sea bream spawns in off-shore spawming grounds mainly in the months of April and May. The eggs begin to hatch as they are being carried to the coastal waters by the warm currents. When the fry reach the shallow coastal waters they enter a period of life near the bottom, where they grow along with the warming of the waters in May and June to fingerlings with a body length of about 5 cm. During this stage of their life cycle they spread out their habitat into areas with a depth of 20 to 30 metres, until September, when the water temperature begins to drop, at which time the fish begin to migrate to deeper waters, where they will enter a hibernation stage. They remain in hibernation in deep water until March of the following year, after which they begin to migrate towards shore in April and May as one year old fish and enter the stock of fish catchable by the coastal fisheries. This cycle of migration from deep waters to shallow waters continues until the fish reach the age of three. At the age of four, however, they migrate to the spawning grounds in off-shore waters.

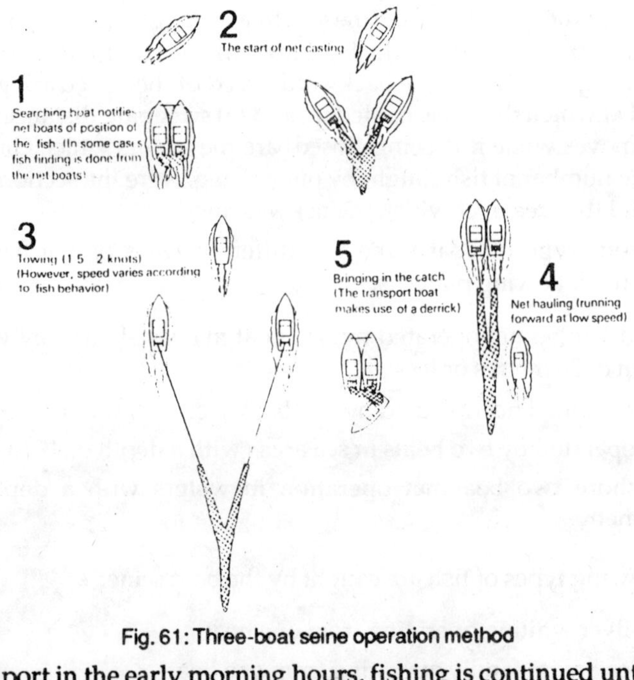

Fig. 61: Three-boat seine operation method

Leaving port in the early morning hours, fishing is continued until just before sunset. First, the buoy is thrown into the water on the port side, after which the boat begins to move in a circle to the left. While circling the left tow rope, net and right tow rope are laid out into the sea. When the net is fully laid the boat heads back in the direction of the buoy at full speed (this takes about 2-3 minutes).

The buoy is pulled up and the end of the tow rope is attached to the drum. The boat then moves forward until the slack in the tow rope is taken up, at which point the drum is started and the net hauling begins. When the bridle of the net reaches the surface, the drum is stopped and the net is hauled in by hand while it is brought around to the starboard stern area of the boat where the bag net will be brought up by hand.

After the catch is brought in it is sorted on the stern deck and the quality fish, such as, red sea bream are put in the live fish hold and the rest of the catch in ice storage.

6

PURSE SEINE

Seiners

Surround nets of many types have been used the world over to catch surface swimming fish. Ring nets, lampara nets and tuck seines are the best known of the numerous kinds of surround net that preceded the best known and most efficient one of all—the purse seine. Modern purse seines are so large and so efficient, their use has seriously depleted stocks of herring, anchovy, pilchard and mackerel in different parts of the world.

There are differences in size, shape and use between purse seine and ring net or lampara. But the chief distinguishing feature is that a purse seine has purse rings fitted all the way round the foot. A ring net or lampara may have a few rings or none at all, but a purse net must be rigged with a continuous series of rings so that it may be "pursed" or closed in a single winching operation.

The first recorded use of the purse seine was off Rhode Island, U.S.A. around 1840. Purse seines are used now the world over, in tropical, sub-tropical, temperate and arctic seas. They are used from small canoes and open boats to large refrigerated vessels of 80 metres length with up to 6000 h.p. engines.

Common Features

Some features are common to most pursers. Since they are able to catch large quantities of fish, they need to have fish holds large enough to accommodate the big catches, and equipped with fast and efficient ways of loading and unloading. As the search for and location of fish schools comprises a large part of a purse seiner's work, the vessel requires a considerable amount of equipments for fish detection and communication with other-searching vessels. In areas, where visible sighting is possible, a crow's nest or look-out may be constructed on the foremast.

Skiffs

Since purse seiners must be able to manoeuver in close proximity to the net without fouling the propeller, most of them require the assistance of a powered skiff or small tow boat. Larger seiners are fitted with bow thrusters, and sometimes both bow and stern thrusters to provide them with side propulsion. Skiffs vary a lot in size, design and power. Most of them are beamy vessels with powerful high speed engines. Small purse seiners may manage, however, with an out-board engine powered skiff.

Net Haulers

Net hauling equipment is another vital part of a seiner's outfit. The chief type of net hauler used on purse seiners is the hydraulic power block. It greatly reduces the time and labour involved in taking in the net after it has been pursed.

In addition to power blocks, net drums are also used to pull in seine nets. Its chief advantage lies in that it greatly reduces the number of crews on a vessel. On the larger herring boats, a power block or a transport roller is used in addition to the drum boats which use drum only are fitted with a stern roller and a pair of guide on rollers for the net.

Two main types of block in use are the sheave and the triple-roller. Sheave blocks are usually hung from the main derrick, but they may also be fixed to the side of the wheel house on a vessel with aft bridge construction. On small boats the block may be mounted on a pedestal. Triple-roller blocks are used in vessels with bridge aft construction. They are fixed on the starboard side of the superstructure. Vessels with this layout also use a transport roller over the net bin aft to lead the net on from the power block.

Winches

The purse seine winch is similar to a trawl winch except drums are of double capacity. Parallel drum winches are common on vessels with forward superstructure. The need for a winch drum of double warp capacity arises when setting the gear. When pursing, the lines are hauled in on seperate drums.

Purse seine gallows are placed forward of midships and so that the warps lead on to the winch. Most purse gallows can be made to lean out over the rail when fishing, by means of removable bolts in the support.

Fishing Side

European and Scandinavian purse seiners, like most trawlers, fish only on the starboard side. Most tuna seiners, however, fish to port as do some salmon vessels and sardine seiners in some Asian countries.

Whether a vessel will fish on starboard or port side depend on her mano-euverability and particularly, the direction of the rotation of the screw and also on skipper's preference.

Bridge Equipment

Many small purse seiners fish successfully without any bridge equipment. But for successful fishing with powered decked vessel, besides magnetic compass, the chief instruments necessary are for fish detection and for communication.

Communication equipment comprises long and short distance radio-telephones.

For long distance SSB (single side band) radio-telephones and for short distance transmission vhf (very high frequency) radios are most popular.

For fish detection the paper-recording echosounder is in widespread use. Low frequency (10 to 40 kHz) sounders are best for herring, sardine or demersal fish. High frequency (80 to 200 kHz) sounders are better for locating mackerel, tuna or squid and other species not having an air bladder.

Horizontal searching is accomplished by means of sonars. Sonars are expensive instruments requiring skill and experience in their use. Sonars are best installed with the transducer placed about one-third of the vessel's length from the stern and can sweep an area of 30, 60 or 90 degrees. Some can scan constantly all 360 degrees around the vessel. Presentation can be on paper or on a p.p.i. screen. Some Norwegian seiners have small hand operated sonars installed in their seine skiffs.

Fish Handling

Fish handling equipment on a purser includes loading and unloading devices. Loading of fish from the net is done by means of a dip net or brailer from a swinging derrick. Fish are dropped out of the brailer into the hold or into fish pounds on deck and from there through bunker leads into the fishroom below.

Specially designed fish pumps, without damaging fish, are used by all the large modern pursers.

The common method is to put the purse seine catch in bulk in fish hold Ice should be sprinkled through the catch and must be taken immediately to market for quick sale. A better way of handling the catch is to box it all in the hold. A shovel of ice is to be put in each box along with the fish. Using chilled or refrigerated sea water tanks (50-60 per cent fish and 40-50 per cent sea water and ice) is another way of maintaining good quality of the catch.

Ocean going tuna purse seiners freeze all of their catch, after bleed washing and putting into a brine freezing tank. After freezing tuna are stacked in a refrigerated hold with minus 20°C.

Large catches of perishable fish must be unloaded quickly at the port. Unloading basket by basket, either by hand or using a landing derrick, the fastest speed a crew can maintain is about one tonne every 7 to 10 minutes. Unloading boxes, up to ten boxes using ship's derrick at a time, with the help of pallets and slings, a speed of one tonne every 5 minutes can be maintained.

From refrigerated sea water tanks, fish are usually brailed out with a dip net. The net is emptied into a pound on the pier or truck from where the fish are scooped into boxes already iced. Frozen fish are simply unloaded by menas of a net sling and put into refrigerated store.

Stability

Stability, which is of over-riding importance on all fishing craft, must be doubly checked on a purse seiner. The weight of the net on the power block, the large catches loaded and the weight of all the deck equipments (including skiff), combine to reduce the stability of a purse seiner. It must, therefore, be carefully checked for all conditions.

Some traditional purse seiners in the North Atlantic have been able to improve their vessel's stability by having a shelterdeck, which increase the freeboard and freeboard is a vital component in the overall stability picture.

Purse Seiner Layout

Four basic types of vessels engaged in purse seining are being considered for their relative merits.

No particular layout is "efficient" in its own right. It must be made efficient by wise and careful positioning of every piece of equipmnt and machinery including warp leads, fish pounds, gallows, derricks, winch, power block, fish pump and water separator. Every aspect of the operation must be studied and allowed for when laying out the deck.

American West Coast and Tuna Seiner Types

In the great Peruvian fishery for anchovy, in the Pacific herring fishery and in the Californian tuna fishery large and small purse seiners have the same basic layout. The following layout is used in Spanish purse seiners and in French tuna seiners. The wheel house is positioned forward leaving a clear working deck behind. A power block is slung from the end of a derrick rigged to the main mast aft of the superstructure. The winch is placed on deck just aft of the mast, and forward of the fish hold hatch. The net is laid in a large bin or pound at the stern and the gallows placed at the rail level with the winch, on either the port or starboard side. A pin or rod in the rail aft holds the purse rings in order and when extended over the rail allows them to slide off one by one as the net is being set. The skiff is usually laid on top of the net though some seiners carry the skiff on the side deck opposite to the one with the gallows.

All of these vessels have engine rooms forward, fish hold midships and fuel or fuel and water ballast tanks aft. Accommodation is usually above decks in the lower portion of the superstructure, which houses galley, mess deck and cabins. This design is suitable for tropical and sub-tropical areas.

Drum Seiners

Drum seining was developed mainly for small vessels fishing for salmon in river estuaries. The vessel has the same basic layout as west coast pursers with the exception that a drum is used instead of a power block. The drum seining method is easy and efficient. The net is set from the drum after pursing. The purse line is usually made of rope and a simple twin gipsy capstan may be used as a seine winch.

Setting in restricted waters and river channels the vessels encounter strong currents and are often in danger of striking submerged rocks or sandbanks. Both to tow the gear clear of obstacles and to drive the fish back into the body of the net, the seine may be towed a distnace before pursing. The main vessel takes most of the weight of the gear and the skiff takes the first end of the net. With both vessels towing, the large net can be pulled to deeper water and the fish swept back away from the opening. The net may then be pursed in safety with less chance of losing the fish or damaging the gear or the vessel. Except when there is a need to tow the net, most drum seiners set and haul the net without the aid of a skiff.

After pursing, the rings are held in a "hairpin" hook and the purse line allowed to lie slack. The drum is then started and the net is wound on until the fish are driven up in the bunt. Net, rings and purse line all go on to the drum. The net comes in over a roller on the stern. Between the roller and the drum there is a "guide-on" pair of rollers which move from side to side slowly so that the seine is wound evenly on the whole width of the drum.

Scandinavian and West European Pursers

For the protection of crews on deck from heavy seas and adopting other fishing methods, e.g. bottom trawling, long lining, Danish seining for part of the year, the layout of purse seiners in Scandinavian and West European countries must be suited to dual purpose or multi-purpose fishing.

Most such vessels have their engine room and bridge aft. A whaleback is fitted forward and on the larger vessels this is replaced with a full-length shelterdeck. All these crafts are rigged for starboard side pursing.

The power block is fitted to the side of the wheelhouse or casing, just aft of the bridge. All the larger boats have a transport roller fitted on a boom or crane so it can swing over the net bin. The winch is usually placed on the foredeck. Small boats have the winch lying fore and aft straight across from the purse gallows, or athwartships well forward under the whaleback. On larger vessels it may lie at an angle on the port side of the foredeck. Whichever way it is arranged, the warps must be able to run both to the purse gallows and to the trawl gallows. Most of these vessels fish over the stern when trawling and the towing blocks are hung from a gantry or gallows extension aft on each quarter.

Because of large space required on the foredeck for the fish pump, water separator and refrigerated sea water tank hatches, many of these boats are now putting the winch on top of the whaleback or shelter-deck, now common in purse seiners over 30 metres overall length. For stern trawling they are fitted with net drum aft either above or below the purse seine net bin. On some boats the trawl net drum lies to the port side of the seine bin. These vessels may switch from seine to trawl or *vice versa* while still at sea. On the main deck level the superstruture extends all the way across at the port side leaving only the star-board side deck clear from forward to aft.

Small Scandinavian pursers of around 25 metres length may have a whaleback extending from forward to the bridge aft on one side leaving the other side clear for fishing. This semi-shelter deck provides both shade and shelter for the crew and the fish deck. By positioning the winch at an angle the warps can be made to run either to the purse gallows, which should be positioned well forward, or to the trawl blocks aft without any sharp bends.

Boats under 30 metres length carry a long derrick on the foremast. While at sea the end of this derrick rests on the top of the wheelhouse, which is used for fish landing. Using this landing derrick, a skilled crew can hoist boxes or baskets of fish from inside the hold up on to the pier in a matter of seconds.

Small Purse Seiners

Almost any artisanal fishing boat can be converted for use in purse seining. Canoes, out trigger boats and small decked vessels are used in various parts of the South East Asian region to handle small seine successfully. In improving traditional craft or in designing new ones to be built, the following modifications need be considered:

(i) Insulation and fitting out of the fish hold.
(ii) Design and installation of a more compact efficient capstan.
(iii) Design and installation of a simple powered sheave or roller for net hauling.
(iv) Installation of an eco sounder and fish attraction lamps.
(v) Installation of adequate pumps for bilge and deck wash.
(vi) Development of faster method of landing the fish.

The method of propulsion be first decided for considering capstans and winches. If marine service engineers are available in the port, hydraulic drives can be thought of. In case of mechanical drives, it may be arranged off the forward end of the main engine, or more directly from an auxiliary engine on deck. By installing auxiliary engine of reasonable size, not only the capstan, but also the marine pumps and generator may be driven independent of the main engine.

If a main engine drive is installed, it is preferable to have a clutch fitted at the power take-off. Using a belt-drive the winch may be placed in front of the wheel house above the engine room, or on the foredeck above the forward fish hold bulkhead.

Good capstans can be manufactured locally by importing the main component, the worm gear.

Simplest eco-sounder, with a portable receiver and the transducer fitted at the end of a length of pipe can be installed. All parts of the instrument should be kept clean, dry, secured and free from damage by working gear or moving parts.

A mast, derrick and proper gallows ought to be fitted whenever possible without compromising the stability of the vessel. They can make the work much easier. It is better to have two derricks, one aft and one forward. The gallows should be placed well forward and be high enough so that the rings can be raised above the rail when pursing.

Two Boat Purse Seiners

The two-boat system is used in some of the Japanese and Korean sardine and tuna fisheries and also in American menhaden fishery. The use of two boats to set and haul the net speeds the whole operation. Fast swimming schools of fish can be surrounded quickly and the net pursed and hauled in a short time.

In the menhaden fishery small boats are used to set and haul the net. They do not handle the catch at all. A large mothership takes all the fish on board and even carries the seine boats to and from the fishing grounds. The twin vessels measure 11 × 2.3 metres and have engines of 100 to 200 h.p. The boats are made of steel or aluminium alloy of single chine construction. Power blocks are fitted to the modern menhaden boats. The seine net is stacked half on each vessel and the boats are connected by a bow line. As they approach the school of menhaden, the bow line is released and each boat sets its half of the net in a semi circle. Each boat is equipped with a winch, but usually only one vessel does the pursing. Once the fish are drived up in the bunt, the mother vessel comes alongside and pumps the catch out.

Japanese two-boat purse seiners are mostly small vessels, (10-30 tonnes). Tuna pair seiners measures over 100 tonnes and have engines of 500 h.p. or more. Their method of setting is similar to that of menhaden boats except that each boat pulls in a separate purse line. The purse wire is in two parts each extending from the wing end to the central bunt. The power blocks used are large Vee shaped reels situated on the stern or quarter of each boat.

Purse Seine Net Design

Factors for Fishing Gear Design

Fishing net keeps its shape under the upward force (buoyancy) of the floats and downward weight of sinkers and the net even when a current is acting on the net.

A buoy or a float should have low specific gravity, sufficiently water tight, cheap and easily obtainable. Cork, pine bamboo floats have been widely used. Synthetic floats, glass floats are now in use instead of natural materials. The floats used in deeper waters should be able to withstand great pressure.

The buoyancy of float is expressed as the volume of float minus its weight in air. The buoyancies of various types of floats are:

Material	*Specific gravity*	*Buoyancy per litre (gm.)*	*Buoyancy per kg. wt in air*
Cork	0.175	825	4.71
Paulownia	0.294	706	2.40
Bamboo	0.500	500	1.00
Pine	0.598	402	0.67
Vinyl/Sponge (soft)	0.099	901	9.10
Vinyl/Sponge (hard)	0.129	871	6.75
Synthetic rubber/sponge	0.243	757	3.03
Synthetic cork	0.294	706	2.40
Ebonite	0.375	625	1.66
Vinyl pipe	0.379	621	1.64
Glass/15 cm. dia	0.348	652	1.87
Glass/30 cm. dia	0.244	756	3.10

Lead, iron, brass, stone, brick, concrete are commonly used as sinkers. The sinking power is determined as the weight of the sinker minus volume of sinker in air. The sinking power of the various sinkers are:

Material	*Specific gravity*	*Sinking power per litre (kg.)*	*Sinking power per kg. wt. in air*
Lead	11.35	10.35	0.912
Iron	7.21–7.85	6.21–6.83	0.861-0.872
Brass	7.82	6.62	0.872
Glass	2.70	1.70	0.630
Stone	2.60–2.70	1.60–1.70	0.615–0.630
Brick	1.90	0.90	0.474
Sand	1.80	0.80	0.444
Soil	1.50	0.50	0.333
Porcelain	1.72–2.13	0.72–1.13	0.420–0.530
Concrete	3.00–3.15	2.00–2.15	0.666–0.682

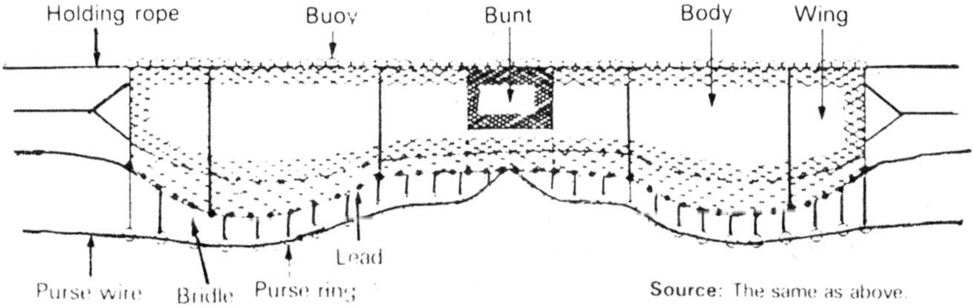

Source: The same as above.

Fig. 62: Purse seine net

Calculation of Twine Weight

The weight of twine for different materials can be calculated as:

Nylon twine	– No. of yarn given × 3 × 0.9	– gives wt in g. per 100 m. length
Vinylon twine	– No. of yarn given × 4 × 0.9	– gives wt. in g. per 100 m. length
Cotton twine	– No. of yarn given × 4	– gives wt. in g. per 100 m. length

Hanging coefficient is expressed as length of the line to which the webbing is hung, in percentage of the total stretched webbing. It is also calculated as the amount of excess or loose webbing (total webbing stretched minus line length) as a percentage of the total webbing.

Purse Seine

Modern purse seine, a sea fishing gear to catch pelagic fishes is classified into two major groups; surrounding nets with bag net and those without bag net. The second group can be subdivided into the nets without purse line (including one boat and the two boat type) and these with purse line.

Surrounding Net with Bag Net

The Japanese semi-surrounding net is a typical example. The bag net is attached to a pair of wing nets. After a boat with luring lamps has attracted a fish school, netting boats towed by other small boat set the net around the fish school.

Surrounding Net without Purse Line

The lampara type net belongs to this category. It has neither purse rings nor purse line in the lower part of the net. This type of gear is operated either by one or two boats.

Surrounding Net with Purse Line

There are various purse seines according to the species of fish aimed at. The net is set around a fish school and then the purse line is quickly pulled on to the boat, pursing the bottom margin of the net.

In one-boat purse seine, the seiner holding one end of the net, surrounds a fish school laying down the net at full speed in such a way as to block the passage of fish school while a skiff is standing by holding the other end of the net. After completion of the setting of the net, the seiner receives the other end of the purse line and bridle from the skiff, and the net is hauled onto the deck as soon as pursing operation is completed.

The purse line or the wire rope for pursing bottom of the net are attached to the sinkers in almost the same manner as in the one-boat purse seine. But the fishing operation is made by two boats. However, the purse seines of one-boat and two-boat are of different construction; the twine size, the mesh size, the length and depth of the net and the ratio of the length to depth. Sardine, mackerel, horse mackerel skip jack and tuna are the main objectives of these purse seiners.

Conditions for Purse Seining

In mackerel, horse mackerel purse seine, operated by one boat, the netting twine for cod end ranges No. 21 to 24 (no. of yarns) and the mesh size 5 to 6 cm. in stretched condition. While in a sardine and small horse mackerel purse seine, the cod end twine ranges from No. 13 to 24 and 3.3 cm. stretched mesh. In two–boat horse mackerel and mackerel purse seine, No. 18 to 21 twine is used for cod end and the mesh size is from 5 to 6 cm. In sardine purse seine No. 6 twine is used in cod end with 1.7 cm. mesh. Seiners operating in bays and inlets have 1.1 cm. mesh in No. 4 to 6 twine.

In Thai purse seiners, No. 4 to 6 twine with 3.3 cm. stretched mesh is used in webbings; while for catching Indo-Pacific mackerel, bonito a twine size between number 9 to 12 with stretched mesh size between 3.8 and 4.5 cm. are used.

To prevent the float line from sinking below the surface during pursing operation, the purse line has 2 to 4 times of buoyancy against the total sinking power. To prevent escape of fish school from the lower part of the wings, the net must have a fast sinking speed. Average sinking velocity of the purse seine are in the range of 11 to 17 cm. per second, the highest and lowest sinking velocities being

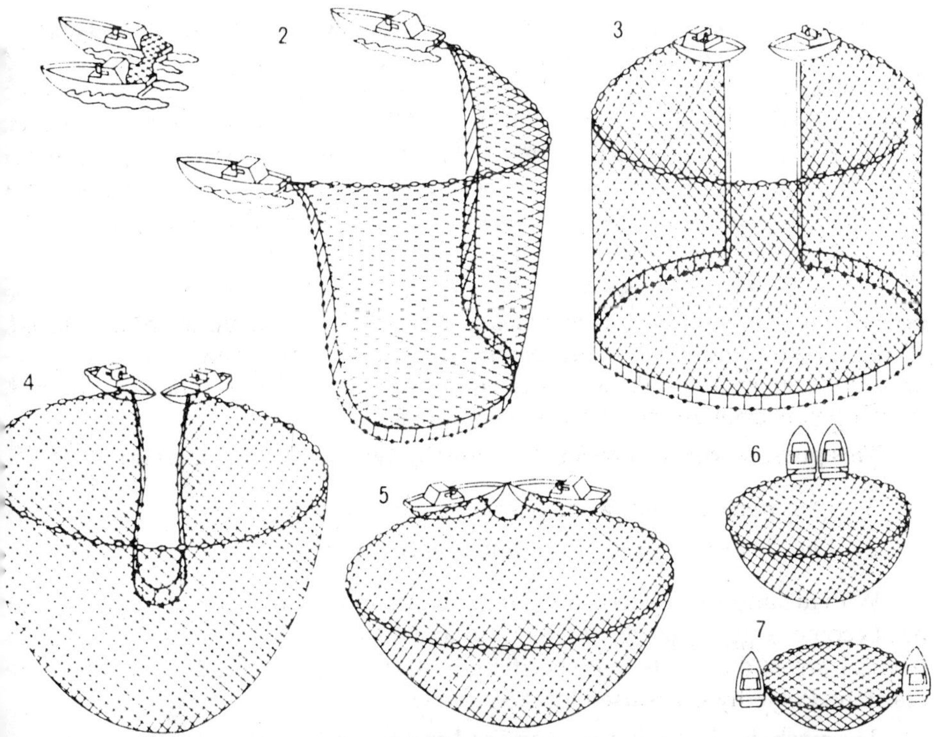

Fig. 63: Purse seining method

27.5 cm./sec and 4.2 cm/sec. respectively. The average sinking power per metre ranges from 1.1 to 2.8 kg/m. Some herring purse seiner in Iceland use heavy sinkers ranging from 9.2 to 9.9 Kg./m.

Tetoron or Vinylon are used for the webbings of purse seine due to their large specific gravity. The specific gravity of nylon is increased by treating it with resin.

Sinking velocity of the net can be increased by increasing the sinker. But even if the weight of sinkers are increased to two or three times of their original weight the ratio of sinking velocity increases only by 1.35 and 1.5 times respectively. With an increased number of sinkers, floats must also be increased to prevent the float line from sinking below the surface during the fishing operation, making the net more bulky for handling.

Purse Seining Method

Modern one-boat purse seiners originated from two-boat purse seines. The two-boat seiners consisted of an engine powered mother boat and two non-powered boats. When a fish school is detected, one half of the crew members move into one non-powered boat and the other half on the other non-powered boat and start to shoot the net. The operation is carried out by these two non-powered boats under the command of master fisherman on the mother boat. As all the work on board has to be done by manpower, many crew members are needed. The fishing gear used is also small in size, because it has to be handled by manpower.

Modern one-boat purse-seines, on the other hand, consist of only an engine powered boat with purse line winches so that manpower requirement is reduced. With the help of the winches, the seiners can use large size fishing gear. As the one-boat purse seiner is larger is size and is equipped with modern machinery on board, it can operate even under rough sea conditions.

This purse seiners can go for two different types of fishing operations:

1. One week's cruise, when operating day and night.
2. One day's cruise, when operating only at night.

With the improvement of the gear materials and machinery on board, the gear has become large in size. Nylon 6 or 7 filaments have been developed for use in tropical areas, due to its low specific gravity (1.14–1.51) and high temperature resistant. Polyethylene is used for the selvedge.

The mesh size for the main net and cod end part is 2.5 cm in stretched condition, and for the twine of the main net nylon 12-ply and for the cod end nylon 150-ply are used. Polyethylene 380 deniers 12-ply twine is used for the selvedge net. The hanging coefficient is 30 per cent on the float and sinker sides.

Synthetic rubber, flat type floats of 310 g. buoyancy each are used for the main net part, while floats of 380 g. buoyancy each are attached to the cod end. For additional buoyancy, round floats with 1000 g. buoyancy each are used around the cod end.The weight of lead sinkers is only 0.5 kg. per metre of the sinker line.

Iron or brass rings of 0.5 kg. weight each with 12.5 cm. outer diameter and

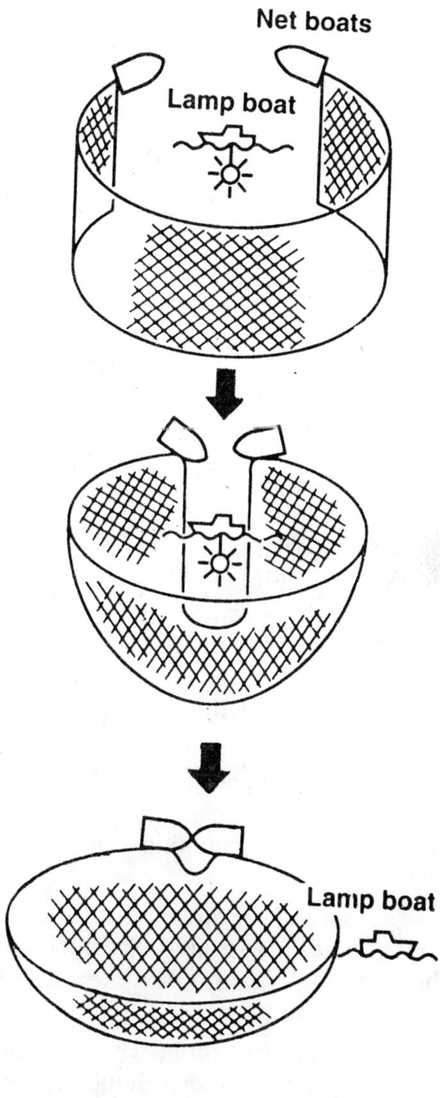

Fig. 64: Luring fish in purse seine with lamps

10.0 cm. inner diameter is used for the purse line. A polyethylene cross rope of 38 mm. diameter is used as rope for the purse line.

Polyethylene ropes of 6 to 8 mm. diameter is used for float and sinker lines. During hauling, the float line is pulled towards the boat by using an auxiliary rope, which is set parallel to the float line through the ring fixed to the float line so that the auxiliary rope can be moved freely and the rope can be pulled through a winch.

Fishing Method

There are two methods of detecting fish schools; (i) traditional visual observation in which a master fisherman during evening and morning detects fish schools by observing a change in water colour and foam appearing on the surface, and scouts fish schools at night by observing plankton luminescence caused by swimming fish schools and (ii) by using fish finders.

Fishes are lured by two ways, one in night and the other for day operation. Acetylene gas lamps are effective to lure fish and are utilised when no fish school can be detected or the school detected are small and not dense.

The fish luring device by acetylene gas lamps consists of 10 to 15 rafts, on each of which two or four acetylene gas lamps are installed. According to the master fisherman's judgement, the seiner is anchored at a place where fishes are expected to be attracted. The lamps are lit on and the rafts are allowed to drift away one by one from the port side of the boat. The rafts are connected with rope to each other keeping a distance of 100 metres and finally connected to the boat. After 4 to 6 hours, necessary to attract fishes, all the rafts are gathered together and moved away as far as 150 metres from the purse seiner. Anchor is weighed and the crew starts shooting the net from a suitable place against the current and wind in such a manner as to surround the rafts. During shooting and hauling of the purse seine, the rafts are kept at the centre of the circle drawn by the purse seine. After pursing operation, the rafts are moved out of the seine and the lamps are turned off. Two crew members take care of the rafts during the time of operation.

Coconut leaves are set at various depths in water attract fish schools. The device consists of a rope, coconut leaves, a bamboo pole and stones. Rope length is slightly shorter than the depth so that the bamboo pole stands upright above the surface. One or two days before the operation, several sets of this device are set in the fishing ground. When the operation begins, these are gathered together so that all the fish lured are concentrated in one position. Shooting is made so as to surround the fish with the net.

After detecting a fish school and the direction of its migration, the purse seine, kept in starboard side, is shot so as to surround the fish school. Shooting is made in such a manner that the seiner can return exactly to the initial position within 4 minutes, when the shooting is completed.

An end of the float line is pulled up on the boat by hook, and immediately an end of the purse line, which has been connected with the float line is set to one winch. Another end of the purse line is also set to the other winch. Then the purse line continues to be pulled until the pursing operation is completed within 20 minutes.

Hauling of the net is made by manpower from the starboard and port sides simutaneously. When the net is hauled up keeping the cod end in the water, hauling from the port side is stopped and the cod end is moved to the starboard side within 40 minutes to scoop out the catches.

During scooping the fish, the upper part of the cod end is hauled above the surface to prevent the fish from escaping. After hauling has been completed the fish are scooped into the fish hold by the scoope net witin 40 minutes.

After the operation, the fishing gear is kept ready at the starboard side for the next operation within 40 minutes. Thus two and half hours are required for one operation.

Comparison between Purse Seine Fishing Operation

While there is little difference between the types of fish caught, by the different types of purse seine using one or two boats and those using a purse line or not, fishermen in different regions still choose the type of fish they will catch according to the season.

The 19.9 tonne class boats and larger are used exclusively for purse seine fishing operation, while under 10-tonne class boats can be used not only for purse seine, but also for small trawl net, boat seine. A large-size boat is operated with the primary aim of increasing the amount of catch, while a small-size boat aims at catching prime fish species, rather than mere increase of whatever catch. A small-size boat is much speedier (19 to 23 knots) than an large-size boat (8 to 9 knots).

Operational details of purse seine fishing boats

Purse-seiner size	*Fleet size*	*No. of crews required for operation*	*No. of fishing days in a year*
39.9-tonne class	Purse seiner–1 Transport boat–2 Lamp boat – 2 Skiff– 1	35	130–150
19.9 -tonne class	–Do–	14	130–150
9.9-tonne class	Purse seiner–1 Transport boat–1 Lamp boat–2	12	120–130
4.9-tonne class	Purse seiner–1 Lamp boat–2	6	120–130

One-boat type is more suitable for off-shore fishing operation. While one-boat type can be operated even on rough sea conditions (wind velocity of up to 15 m./sec) thus increasing the number of fishing days, which save labour and reduce capital and operation expenses; the two-boat type can be operated even in shallow or reefy waters, reducing net casting and hauling time and easy net hauling even in swift currents.

Various two-boat purse seine operations

Purse-seiner size	*Depth of fishing ground in metre*	*Fishing season*	*No. of crews*	*Net size (Buoy side tailored length)*	*Mesh size*	*Catch*
Sardine purse seine under 5-tonne	Under 50 m.	All year (Sept-Dec) Peak	24–30	300–500m.	28–22 mm.	Sardine, Horse mackerel, gizzard shad, black porgy, grunt.
Luring lamp sardine purse seine –Do–	40–80 m.	All year (April-July) Peak	35–40	600 m.	Body–16mm Wing-18mm.	Anchovy horse mackerel, grunt, mackerel, scad
Gizzard shad purse seine –Do–	40–60 m.	Oct. -April (Oct-Dec. Peak)	6–8	100 m.	38 mm.	Gizzard shad
Hemiramphus purse seine –Do–	30–40 m.	Nov. to May (Dec. to Mar. Peak)	6–8	70-240m.	Body-22-20mm Wing–19-18 mm	Hemiramphus
Sardine purse seine 19.7-tonne	under 50m.	June to Nov.	25	500 m	11–10.8 mm.	Anchovy, sardine, mackerel, barracuda
Sardine purse seine 19.9 -tonne	200–300m.	Aug to Nov.	32	200 m.	Body– 11.6–11m Wing-9.8 mm	Sardine, horse-mackerel mackerel

19.9-Tonne Class One Boat Purse Seiner

A. *Main Specification and Equipments*

Hull – F.R.P.

Gross tonnage – 19.9 tonnes

LOA – 16.5 m.

Beam – 4.08 m.

Draft – 1.32 m.

Main Engine – Max. 550 h.p. (rated 150 h.p.)

Fishing gear –

(a) Net hauler – Max. capacity 6 tonnes
Hydraulic pressure – 150 kg./cm^2.

(b) Power block – 1

(c) Side roller – 3

(d) Purse seine winch – 1

(e) Capstan – 1

(f) Purse wire reel – 1

(g) Auxiliary capstan – 2

B. *Transport Boat*

Fish locating equipments

(a) Radar

(b) Fish finder

(c) Colour-display fish finder

(d) Sonar.

C. *Lamp Boat (Maximum –2)*

Fish-luring lamp

(a) On both sides of bridge – 2 kW – 4

(b) Underwater lamp - 2 kW –1

The underwater fish-luring lamp is put under water at a depth of 4 to 5 metres. The source of light must be placed lower than the keel.

D. *Fishing Gear*

1. Holding rope – Nylon 36 mm. dia, 20 m. long.
2. Buoy – Synthetic float, placed about 30 cm. interval.
3. Buoy line– 15 mm. dia, 731 m. long.
4. Warp line– 12 mm. dia wire, 200 m. long.
5. Purse line– 17 mm. dia wire, 1000 m. long.
6. Lead– 560 gm. weight placed about 20 cm. interval.
7. Lead line– 12 mm. dia, 780 m. long.
8. Purse ring– Iron, placed about 8 metres interval.
9. Bunt– Nylon–210 denier, 12 yarns, 23 mm.
10. Body – Nylon – 210 denier, 9 yarns, 23 mm.
11. Wing – Nylon– 210 denier, 9 yarns, 23 mm.
12. Buoy line hanging ratio – 0.75
13. Lead line hanging ratio – 0.80

E. *Facts About Operation*

1. Number of fishing days per year – 130 to 150 days.
2. Daily operation period – 10 hours from sunset till sunrise.
3. Time required to reach fishing grounds – 2.5 hours.
4. Operation time:

   (a) Net casting – 5-6 minutes.

   (b) Purse line hauling – 20 minutes.

   (c) Net hauling – 40 to 50 minutes.

   (d) No. of operations per day – 2 to 3.

F. *Fishing Method*

Considering the weather and sea conditions, the fleet leaves port at such a time, so that it can reach fishing grounds at sunset. On reaching the fishing ground, the purse-seiner and two transport boats begin to seek a school of fish by means of a fish finder over their assigned sea areas, while maintaining mutual telephone communication. The searching is done with special care, sometimes for two to three hours. The master fisherman decides on the timing of net-casting by taking into consideration all essential factors, such as, fish species the size and density of a school and the depth of water where the school has been located.

The lamp boats are then brought to a right position above the school. Fish luring operation starts by lighting all luring lamps.

In the meantime, the purse-seiner begins its net casting operation as directed by the master fisherman.

The holding rope is released from the stern of the purse-seiner with its end being held by the skiff.

The purse-seiner lets out the net from the stern in order of the holding rope, bunt, body, wing warp line, while moving at full speed around the lamp boats. In the case of 19.9-tonne class boat, this operation must be performed against the current. When the operation is finished, the net surrounds the sea area of 731 metres round and 200 metres deep.

When the purse-seiner receives the holding rope and the end of the purse line from the skiff, the holding rope is hauled in and the end of the buoy line is tied to the bow bitt.

Net hauling begins with the wing side first by means of the net, hauler and power block on the sterm, while purse lines (on both ends of the net) are being rolled round the reel on the bow deck to purse the net.

When the pursing is completed, the crew on the bow deck begin to disconnect the short bridles from the purse rings by threes or fours to meet the speed of net hauling, thus allowing the net to be taken on the stern deck. The net is then folded up in good order for the next casting.

The lamp boats stay in the middle of the surrounded sea area until the net is completely pursed. After confirming the finish of net pursing, these boats move to lure the school to come toward the bunt. All lamps are put out when the luring is completed.

When the purse-seiner has hauled in the wing and body, one transport boat ties the buoy line of the bunt to the side. In the meantime, the purse-seiner begins to haul in the net by means of the side roller. Fish are scooped with a big dip net when they come up to the surface of the water.

About three net castings are performed a day. In the case of a big catch, one casting can put an end to the operation of the day.

Characteristics of Two-boat Purse Seine

The fishing season and ground is the same as one-boat purse seine. The boats are operated to catch mainly sardine, anchovy, horse-mackerel, mackerel and yellowtail.

The fishing method is also the same as one boat purse-seine except the following:

1. Operation is usually limited to the day time.
2. Two purse-seiners of the same type share the load of a net with their bows linked.
3. Each purse-seiner is equipped with a purse wire winch and a net hauler.
4. Time needed for each casting and purse wire reeling is reduced by half. This means the number of daily castings is increased, thus resulting in a better catch.

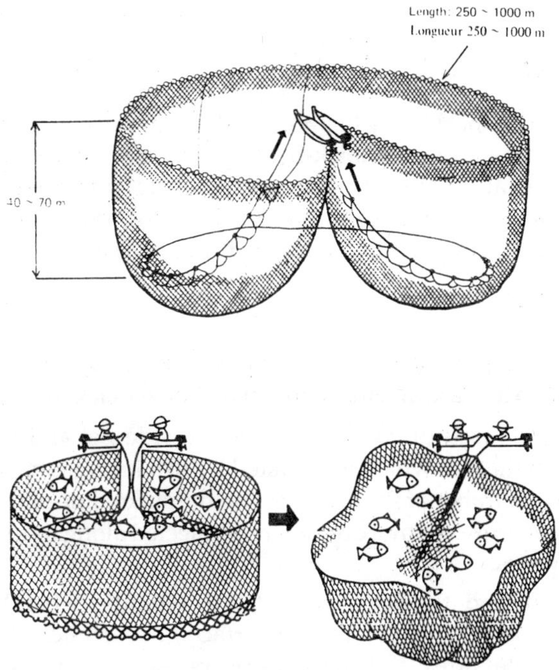

Fig. 65: Purse seine operation (Two boats)

Operation Method

1. Sharing the load of a net and moving forward with their bows linked.
2. Separating from each other and casting a net when a school of fish is located.
3. Surrounding the school.
4. Pursing starts.
5. Net hauling.
6. Net hauling.
7. The bunt is positioned between two boats and fish are scooped with a big dip net.

7

DIP NET

Stick-held Dip Net

The stick-held dip net fishing has been widely operated in Japan since old times. In Thailand, catching squid by stick-held dip net started after 1978. The gear is suitable for small-scale fishing at a low cost. Mackerel pike is commonly caught in the net. The operation is done at night with fish luring lamps to attract the fish; and has become one of the important fishing methods in Thailand.

Outline of the Net

The stick-held dip net is one of the many types of lift nets used for catching pelagic fishes, such as, sardine, mackerel pike, horse mackerel, squid etc.

The net is composed of cod end, main net, side net and selvedge net. It is designed to scoop the fish which is attracted by fish luring lamps or baits. The net is provided with a number of small sinkers on the sinker line of the net and heavier sinkers to which about six or more ropes are connected along the front end of the net for hauling purposes.

At the opposite end, the net is connected to a bundle of bamboo poles to keep the upper part of the net afloat on the sea surface. On each side of the upper part of the net end, some floats are attached for the purpose of preventing the escape of fish during hauling operations.

Two bamboo poles are used for the purpose of stretching out the net during operation. These are taken off from the net when the net is hauled toward the boat. Also a number of rings are attached to boat sides of the net. Through these rings, some ropes are led for the purpose of pulling up the sides of the net immediately to prevent the escape of fish during hauling.

Operation Method

Motorized fishing boats are being used for operation of stick-held dip net. The net is set in fishing grounds where migratory schools of fish are mostly found. The operation is done at either side of the boat in order to make the operation easier and more convenient.

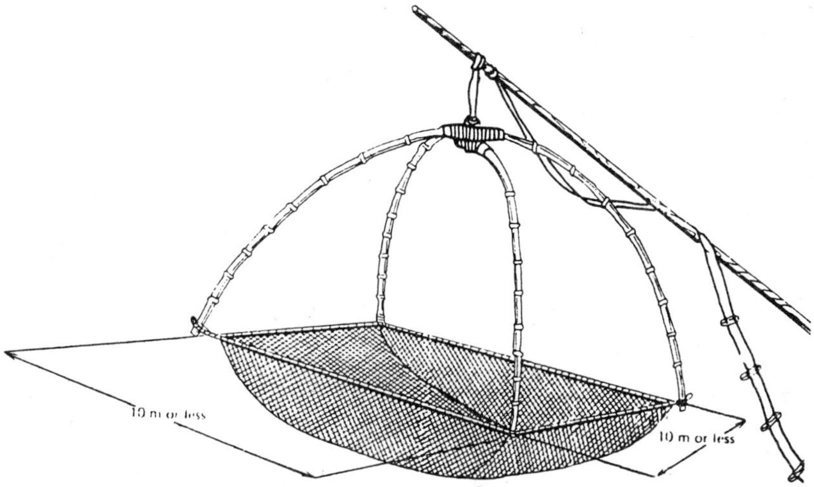

Fig. 66: Stick-held dip net (Pulling type)

The stick-held dip net is mainly operated with fish luring lamps to attract fish during night. When the gear is operated in day, some bait is used to attract fish. With these luring devices, fish are made to concentrate on either side of the boat where the net is set.

Hauling lines are pulled by using the line hauler until the front part of the net is hauled up to the sea surface to prevent the fish from escaping. The bamboo poles are taken off from the net or pulled with the net toward the boat so that it will be easier to scoop the fish being caught. The operation is done 20 to 80 times a day.

Fishing Boat

There is no specific design of the boat, but the boat should have preferably the following characteristics :

(a) Shallow draft	—	The fish should be able to pass underneath the bottom of the boat to the side where the net is set.
(b). Shallow freeboard	—	Shallow freeboard is essential for easy operation.
(c) Resistance against wind	—	The boat should have resistance against wind, to maintain the relative position between the boat and the net during fishing operation.
(d) Fish hold	—	The fish hold should be partitioned into small chambers, so as to keep fish caught at different times in different chambers.

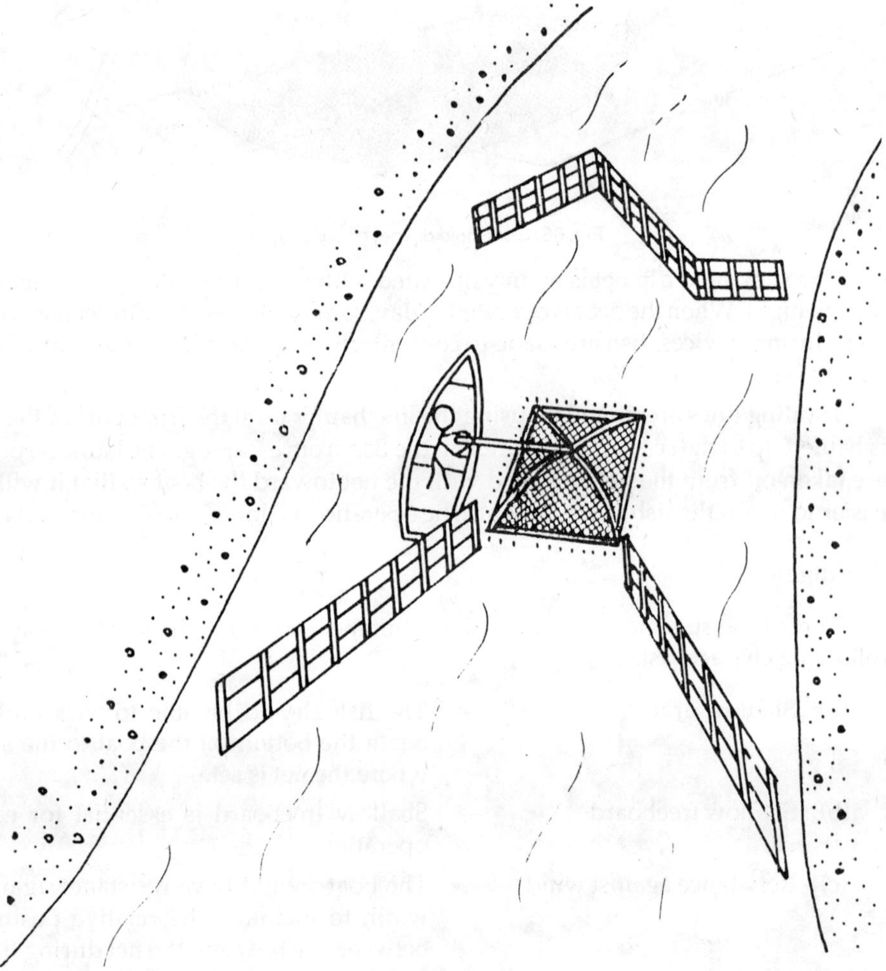

Fig. 67: Operation of dip net

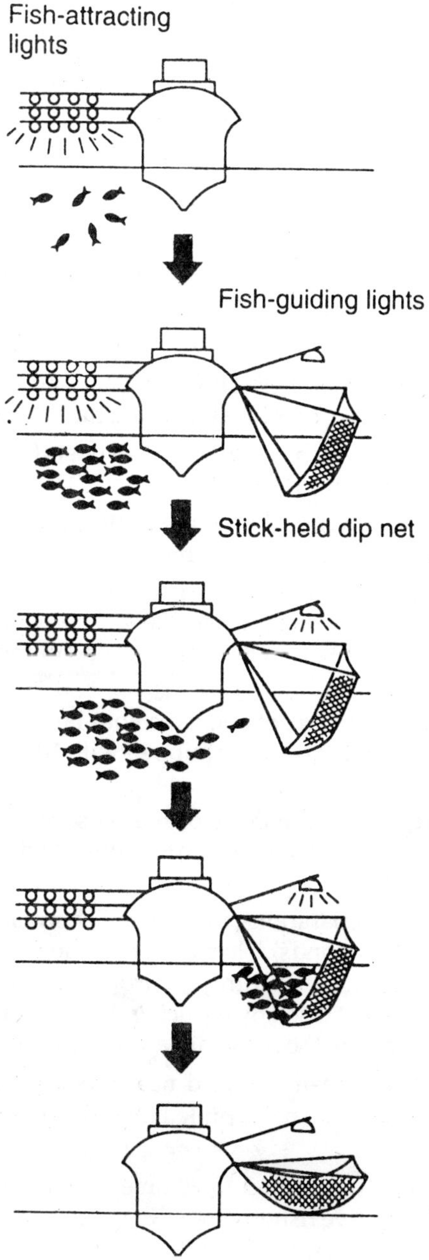

Fig. 68: Dip net operation with fish luring lamp

Fishing Equipments

(i) Fish luring lamp — The success of fishing method depends on fish luring lamps. Fish luring lamps are fitted in 5 to a booms positioned 1.5 to 2 metres above the sea surface.

(ii) Winch and roller — A six-reel winch is provided for hauling the net towards the boat and two auxiliary winches are installed on the boat sides of the stern. To take up the net to the deck, rubber, rod-shaped rollers are provided on the bulwark top.

Squid Stick-held Dip Net

There are two types of stick-held dip net in Thailand. One is used by large boats (15 metres and more), and in order to catch squid, the net is pulled towards the boat.

The other type of net, which is used by small boats (15 metres and less) catches squid by being pushed against the boat.

The nets used in two types of boat are quite similar in construction and are operated in the fishing grounds of 10-20 metres of depth.

Pulling net type—After arriving the fishing ground, the engine of the boat is stopped. One set of the mid-water gill net (300-400 metres long and 14 metres deep) is thrown out to the sea as a sea anchor, and the end of the net is joined to the bow side of the boat.

All the fish luring lamps except one red colour lamp, are switched on to attract the squid around the boat (the boat is equipped with 10 fish luring lamps and 1 red colour lamp).

The squid are attracted around the boat after some time. The net is set under water, then the ends of three lift lines which are connected to the sinker lines are held by three fishermen.

After checking these peparations, the master fisherman switches off all lamps at the same time (within 1-2 seconds). Then one of the four lamps on the starboard side is quickly swiched on. All squid attracted around the boat will then shift to the starboard side where the net is set. By reducing the illumination power of the lamp, which has been switched on, the squid are attracted to come up to the sea surface.

When the squid have been attracted near the sea surface, the net is quickly hauled by three fishermen. The net is pulled to the boat side, then the catch is lifted on board.

The whole operation is repeated 15-20 times in the course of one night fishing. The operation is done by five fishermen.

Pushing Net Type

This type of gear is operated at the same fishing ground and a set of mid-water gill net also serves as a sea anchor.

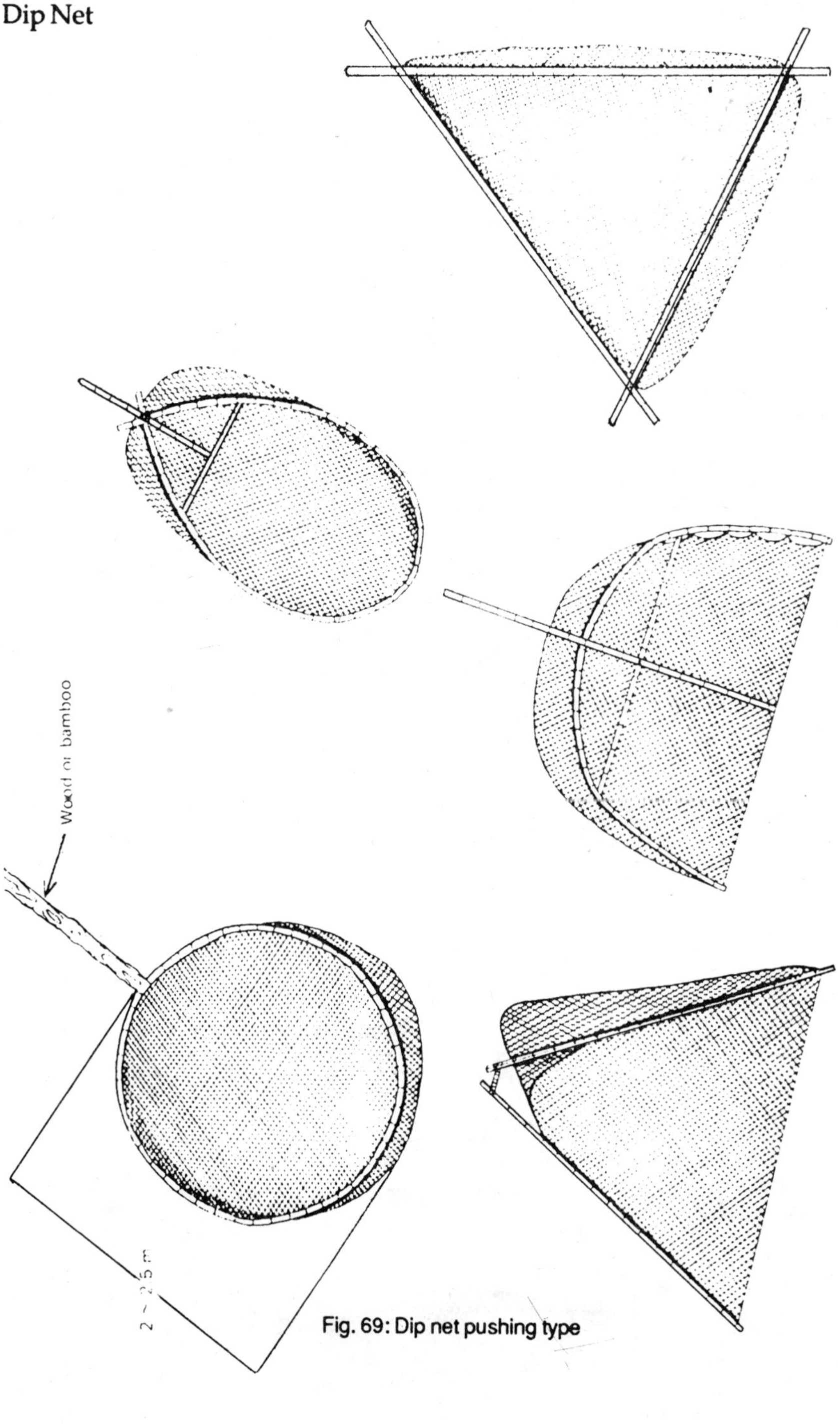

Fig. 69: Dip net pushing type

After arriving at the fishing ground, all fish luring lamps on board are switched on for about 30-60 minutes.

When the squid are attracted around the boat, the port side lamps and bow and stern side lamps are switched off. Four lamps on the starboard side are kept lit, when all squid are attracted to the starboard side by the above procedure the net is set under the water. After that 3 of the 4 lamps on the starboard side are switched off. One lamp is left switched on and its illuminating power is reduced slowly.

When the squid come up to the surface, the net is pushed quickly by two fishermen. The operation is repeated about 20 times in a night by 3 fishermen.

Designing the Net

Before designing the net, it is necessary to know the behaviour of fish. Some fishes, such as mackerel pike, anchovy and squid are easily attracted around the boat by the fish lamp illumination. These fishes are caught with the stick-held dip net using fish luring lamps in the night. On the other hand mackerel, horse mackerel can be attracted around the boat with bait, and thus be caught by this net in day.

The size of the net is decided on the condition of fishing boats, i.e. size of the boat, engine power, number of fishermen, equipments on board etc. Two bamboo poles are used for the purpose of stretching the net during operation. One bamboo pole projects from the bow side and the other one projects from the stern side. So the boat should be long enough to allow the proper distance between the poles and 80 per cent of the boat length is usually considered as the effective length of the boat.

The shape of net, i.e. square, rectangular or trapezoid can be chosen by considering the type of boat and the behaviour of fish.

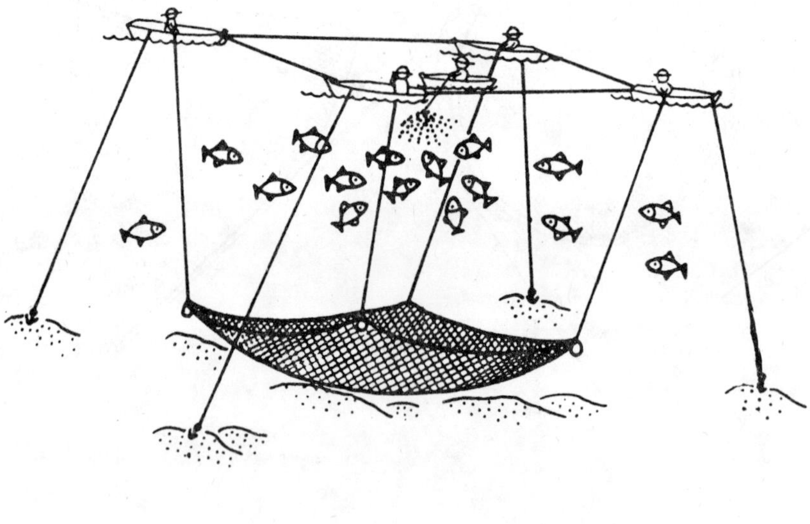

Fig. 70: Lift net using bait to lure fish

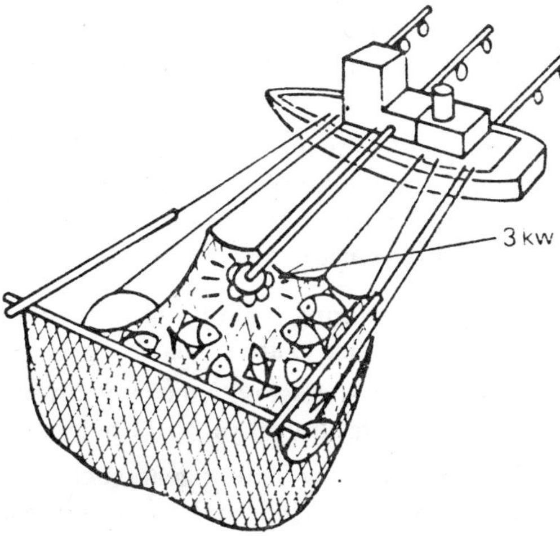

Fig. 71: Lift net using lamp to lure fish

The success of the net depends on whether the net can withstand the influence of sea current and wind and keep a proper required shape during operation. The material of the net and rope should have the following properties:

(i) High breaking strength
(ii) High abrasion resistance
(iii) High density (Specific gravity)
(iv) Low price.

The materials which satisfy these requirements best are the following synthetic fibres:

(i) Nylon (Polyamaid)
(ii) Vinylon (Polyvinyl alcohol)
(iii) Saran (Polyvinylidene chloride) and nylon mix twisted fibre
(iv) Tetrol (Polyester).

The size of the netting twine, rope and accessories should be selected with regard to the size of the net and the fish to be caught.

Lift Net

Lift net is an extremely diversified fishing method that is used all over the world in rivers, lakes, marshes, coastal sea waters and even the open sea, and ranges widely in both scale and operation from small-scale scoop nets, four-armed scoop nets or scarf nets operated by one man without a boat, to stick-held dip nets operated from a motorized boat, two-boat operated lift nets and even four-boat and eight boat large lift net operations.

In this type, a net that has been set out horizontally under water is raised suddenly to catch fish that have either wandered naturally into the water above the net, or have been lured there by fish attracting lights or by scattering bait in the water.

Considering the type of sea area, the lift nets can be divided into two groups: (1) bottom lift nets, which are set on the sea bottom and (2) floating lift nets, which are set in the surface or mid-water. Further, concerning the fish gathering method, the lift net can be divided into three basic types:

(a) Active fish gathering by scattering bait on the water above the net, or at night shining lights on the water to attract fish.

(b) Waiting type—in which the net is set with a bag net in a strategic position for catching fish which migrate naturally into the net with the tidal current.

(c) Methods in which supplementary fishing gear, such as, surrounding gill nets are used to drive schools of fish into the area above the lift net.

Small Scale Hand-operated Nets not Using Boats (Figs. 1–4)

Traditionally these fishing methods including scoop nets, push nets and four-armed scoop nets, are popular in rivers, lakes and marshy areas. They are used for fishing in fresh water lake and catching eel fry for seeding culture operations in estuary.

Large-scale four-armed lift net is used in rivers and shallow sea areas in which fish are caught in a net fixed to a framework made of wood or bamboo. The method is widely used throughout South East Asia. A fishing boat used for transportation between land and the net site.

Lift Net using One-boat (Figs. 5–6)

A method that achieves a high catching rate with a very simple fishing gear. A large-scale scoop net is attached to the bow of a motorized boat to scoop up fish and mysis.

Lift Net using Many Boats (Figs. 7-9)

In this fishing method, a square or fan-shaped net is set out in a fixed position in the water by two or more boats, after which bait is scattered on the water or fish lights are used to attract fish to the area above the net. The net is then pulled up suddenly from the sinker edges into a bag shape and the trapped fish are caught with a scoop net.

Two-boat Lift Net (Fig. 7)

Two boats layout the net in a fixed position. The net is designed so that the tidal current will cause it to blow out into a bag shape. At night, fish lights are used to attract fishes, such as, mackerel, horse-mackerel and sardine for catching. The net used is larger than that of stick-held dip net.

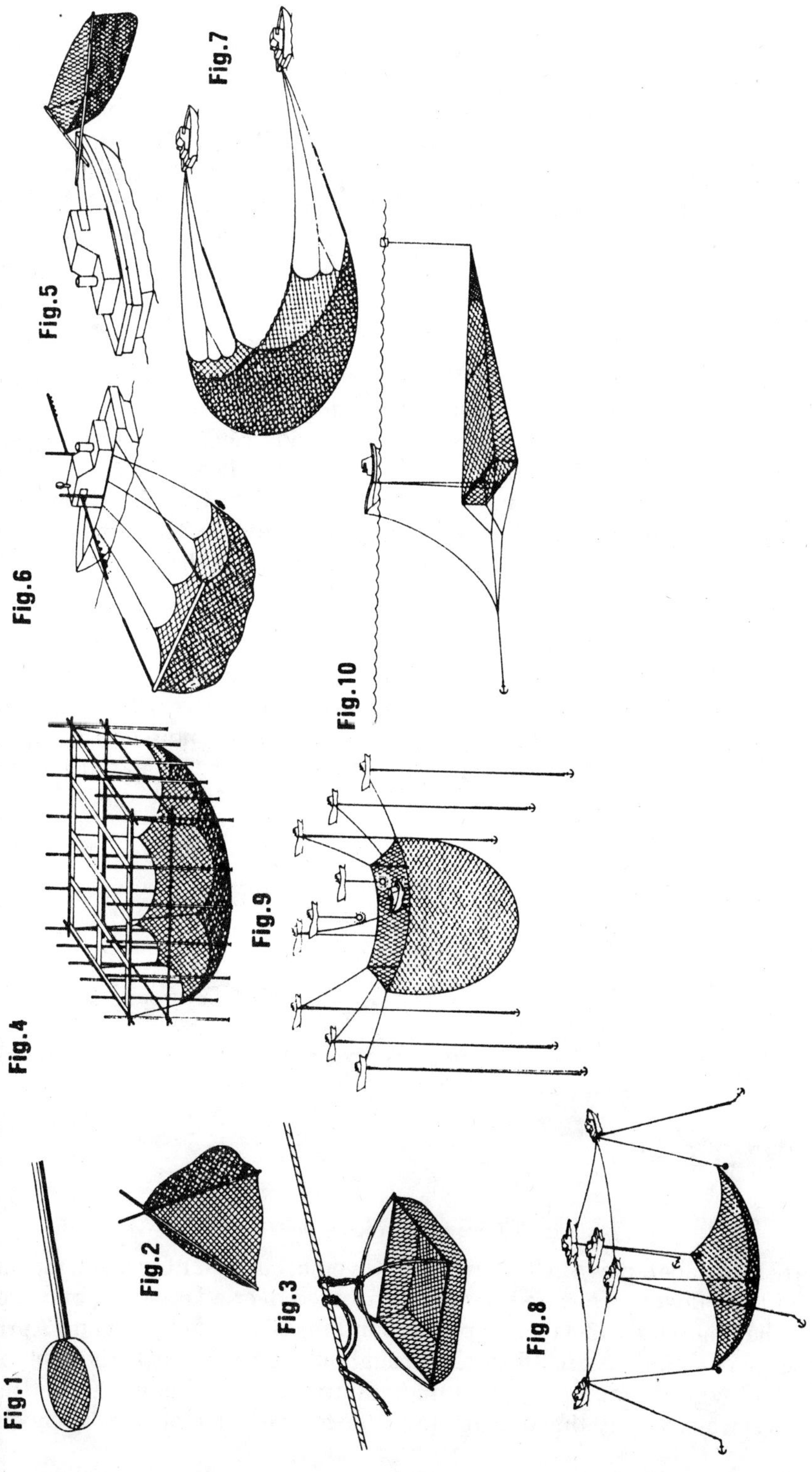

Fig. 72: Different types of lift nets

Four–boat and Eight-boat Lift Net (Figs. 8 and 9)

There are two basic types of net settings, a bottom lift net which is set on the sea bottom, and a floating lift net, which is set off the bottom and held in place by the boats. The later is used primarily in areas with a reefy sea bottom or in areas with a particularly strong tidal flow. In spite of the fact that this method uses a large net, it is not a very effective catching method, when considering, the number of boats and large amount of labour required in its operation. As with two-boat lift net, fish are caught are being lured to the net by scattering the baits or by the use of fish lights.

Specialized Lift Nets (Fig. 10)

This is a waiting-type fishing method that catches fish, which ride the tidal currents into a net shaped like a bag. Usually the net is set with its mouth facing the tidal flow at the mouths of rivers or in shallow tidelands, catching the fish, which ride the current into the net. Sometimes, however, the net is also set vertically in the water with the mouth of the net facing upward, and fish which have been lured into the waters above the net are caught by raising the net suddenly. The fish caught by this method include sardine fry, seabream, croaker and small shrimp.

Drive in Net

This compound fishing method in which a bag shaped net fitted with wing nets or fence nets is set in water and several methods are used to drive fish into the net, including divers, surrounding gill nets, or making threatening noises on the water from fishing boats. This drive-in type net is used for catching "Takabe" (*Labracoglossa argentiventris*) in waters with depths of 20-30 metres in Japan.

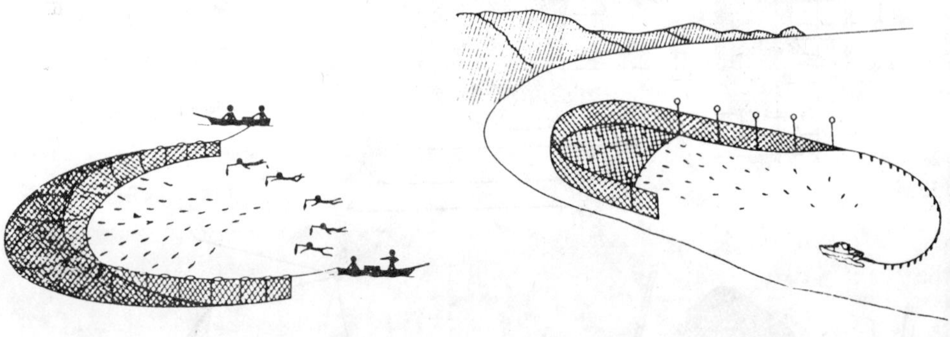

Fig. 73: Drive-in net in operation

The fish (Takabe) spawn in September to November in 18°-21°C water temperatures and their floating eggs and fry are distributed over a large area by water current. The fingerlings (3-4 cm.) appear around the islands in March and April, when the water temperature is 15°-24°C. Throughout their lives the fish is plankton feeders. As they grow they repeat a daily migration pattern of swimming around the off shore reef waters in day-time and moving to the coastal waters at night in schools.

Within one year they reach a maximum size of 22-23 cm, when they attain sexual maturity. The catching age is from 1 to 2 years, with the maximum age reach is believed to be 5-6 years.

Structure of Fishing Gear

The net is constructed of four parts, the four-boat bottom lift net, a surrounding gill net, a divider net and the drive-in net. As for the materials from which the nets are made; the bottom lift net is made of nylon, which cuts through the water easily; the gill net, divider net and drive-in net (with the exception of one segment) are all made of cremona.

Cremona is a short-fibre thread similar in nature to cotton thread. It has a degree of stiffness that makes it easy to work with the hands, which makes it a suitable material for use in nets which must be repaired often, like the gill net, divider net and drive-in net. Cremona also has the advantage that sea weeds do not attach to it as easily as to nylon.

Fishing Method

On arriving at the fishing ground, several young divers get into the water and begin to search around areas of natural reef for schools large enough to catch. A school that looks large enough (700 kg.) is considered proper to call for the beginning of a netting operation. When such a school is located the divers signal to the net boats.

Two work boats layout the surrounding gill nets in a circle closing off at point. In a case like this it is the fishes nature to try to escape by swimming into the tidal current, so the net is always set from the up-tide direction.

Then the divider net boat cuts across the circle made by the surrounding gill net and as it does it lays out the divider net, as quickly as possible. In this way the divider net divides the circle closing off the main part of the school in one hemisphere, and thus reducing the area over which the following catching operation must take place.

Two large net boats are made to wait. Two drive-in net boats use the drive-in net to steadily move the fish in a down-current direction.

At this point twelve or more divers from divers boat, the work boat equipped with compressor and the divider net boat get into the water and drive the fish before them. In this case the bubbles from the divers heads serve to intimidate the fish. When the drive-in net reaches the proximity of one end, the two drive-in net boats leave the rest of the driving process to the divers and processed to the end of surrounding net to join the two large net boats in setting out the lift net.

The surrounding gill net is opened slightly at this end and one corner of the lift net is set into the gap. To this corner of the lift net four 15 kg iron sinkers are attached and it is sent to the bottom. As the lift net is made with an extremely high hang-in ratio, when it receives the tidal current it billows out into a bag shape in the down current direction.

When the school of fish has entered the net, the sinkers are pulled up and the

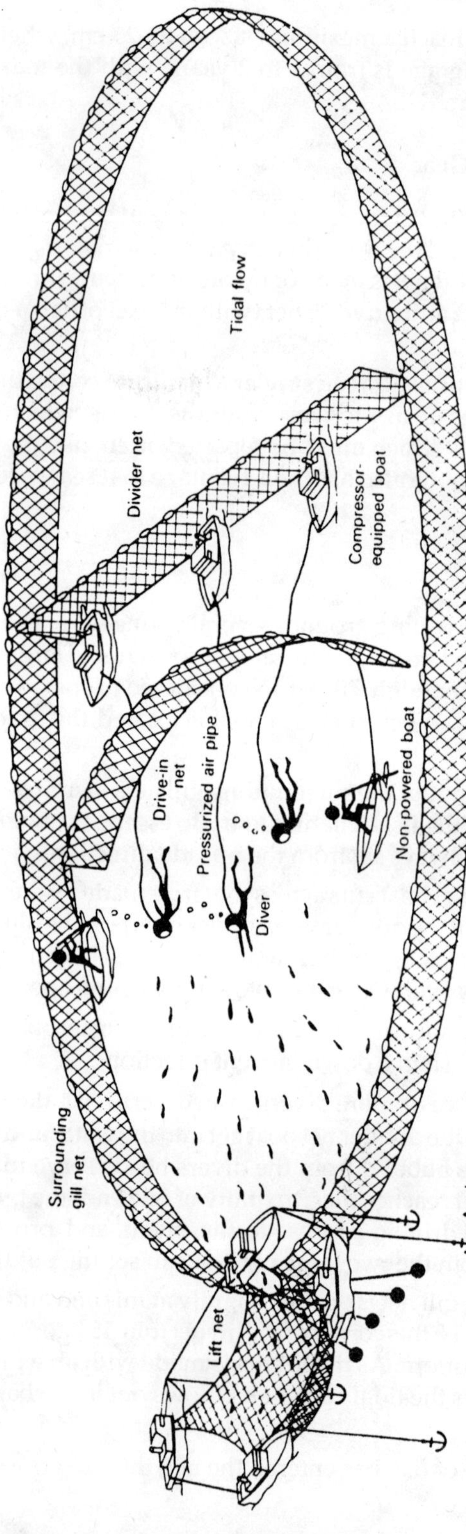

Fig. 74: Surrounding gill net and drive-in net in operation with lift net

four boats at the four corners of the net begin to pull it up in unison and complete the catch operation.

The fishing boats leave the port at 6 o'clock in the morning and continue to fish until evening. Only two or three operations can be made in a day as the entire process take two to three hours as detailed below:

(a) Search for fish schools—50-60 min.

(b) Casting the surrounding gill net —10-20 min.

(c) Driving in the fish with drive-in net and divers—60-90 min.

(d) Rasing the lift net and landing the catch in the boats—30 min.

This fishing method has been specially designed for catching 'takabe' like fishes in large quantities in reefy areas. As 'takabe' inhabit in reefy areas with strong tidal current, it is difficult to use either trawl net or surrounding net. In strong tidal current the surrounding net cannot be set properly and will cause the net to get stuck on the reef, reducing the net's catchability. Stick-held dip net is also not suitable as it is designed for surface fishes and 'takabe' are middle to bottom water fish. Bottom lift net can become caught on the reefs and 'takabe' are not easily attracted by fish lights. This is why the above process of first surrounding the 'takabe' schools with a surrounding gill net, then using a drive-in net and divers to drive the fish to a sandy bottom area where a four-boat type bottom lift net could be set and the fish caught, was developed.

8

GILL NET

Gill Net Fishing

Gill nets are used the world over to catch every kind of fish because the structure of the gear is very simple and one of the cheapest fishing gear.

Gill net consists of a rectangular net with float on its top and sinker on its bottom in order to spread the net in the water.

The gill net is placed in the way of the swimming fish. Attempting to pass through the net, fish are enmeshed or entangled in the net. Generally, spindle-shaped fish, such as, sardine, mackerel, salmon, yellow-tail and skipjack are enmeshed. While crab, prawn, spiny lobster, flat fish, shark, spanish mackerel are entangled in the net. From its catching frequency, the gill net fishing gear has two different sorts of functions—enmeshing and tangling. In designing the modern gill net, therefore, the following points are to be taken into consideration:

(i) How to increase the catching frequency?

(ii) Fish species selectivity of fishing gear.

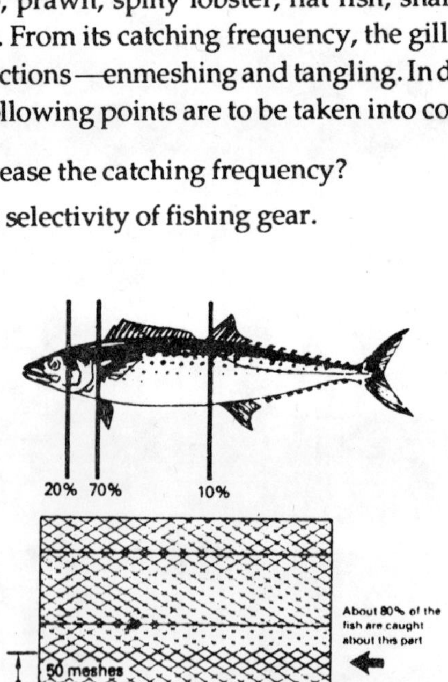

Fig. 75: Catching frequency of gill net

While the mesh size is a crucial factor in catching the fish, it also selects the fish caught by body size or shape. The gill net is by far the most common place for the fish to be caught in the net. The part of the net in which the most fish are caught is the lower fifty meshes of the lower net. Although it is not really a remarkable amount, the greater part of the catch seems to be caught in the central portions of the net rather than the ends. And, very few fish appear to be caught near the seams, where the net has been sewn together.

Therefore, in designing a gill net fishing gear, one must select a type that would serve particular purse best. In addition, much importance must be given to corr-esponding design particulars as well.

Classification of Gill Nets

There are three basic types of gill net depending on the type of fish to be caught and the fishing ground conditions: (a) fixed gill net, (b) drift gill net and (c) encircling gill net. In addition each of these types can be divided into two sub-divisions: (i) those used for surface fishes and middle depth fishes and (ii) those used for catching bottom inhabiting fishes and/or other demersal marine animal, primarily crustaceans.

There are actually six types of nets in use in gill net fishery. Among these, the two nets which have proved to have the best catching capabilities and lend themselves best to a commercialised type of fishing operation are :

Fixed bottom gill net and

Drift gill net for surface and middle-depth fishing.

Classification of gill nets and their representative catch

Net type		*Representative catches*
Fixed gill net	Surface and middle-depth	Sardine, Half beak, Blue sprat, Sandfish
	Bottom	Cod, Alaska pollack, Herring, Mackerel, Shark, Flounder, Flat fish, Sea bream, Lobster, Blue crab Ray, other crabs
Drift gill net	Surface and middle-depth	Salmon, Trout, Horse mackerel, Mackerel, Sardine, Marlin, Spanish mackerel, Flying fish, Ocean perch
	Bottom-gill net	Prawn, Black porgy, Butter fish, Sand smelt, *Scurida undosquamis*
Encircling gill net	Surface, middle-depth and bottom	Spanish mackerel, juvenile yellowtail, Black porgy, Mullet, Grunt.

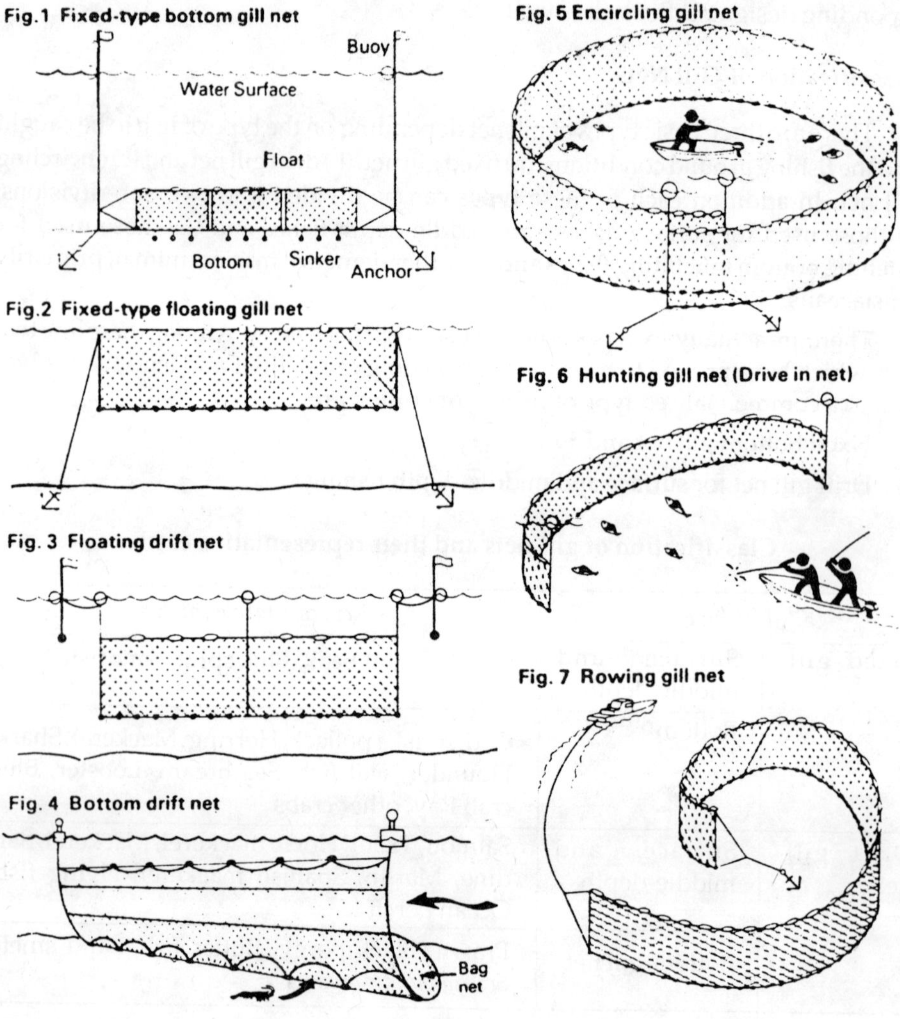

Fig. 76: Various types of gill net

Two other types which must also be mentioned as noteworthy fishing methods are :

Bottom drift gill net for prawn and

Encircling gill net.

The former is an effective fishing method, which makes use of the prawn's behavioural patterns, while the latter is mentioned as a representative type of gill net fishery method that is easy to operate. The variations of the encircling gill net include the "hunting gill net" and "rowing gill net", but these methods are gradually disappearing to the point where they can seldom be found in use today.

Surface gill net catch fish which swim near the surface of the water. Either one end or both ends of the net is fixed to the sea bottom to prevent the net from drifting away by the current or winds while in operation. This gill net is operated on a small scale at the slow current and shallow water area, such as, back bay, coastal water and small strait. Sardine, mackerel and flying fish are caught with this type of net.

Bottom gill net is fixed on the sea bottom with anchor to catch bottom dwellers. The total sinking power of net, ropes and sinker is larger than the buoyancy of float.

Trammel net which consist of two large mesh outside net and one small mesh inside net is a kind of bottom gill net.

Drift gill net drifts with tidal current or wind and catches the fish and does not require an anchor. This type of gill net is most suitable for the pelagic fishery on a large scale, such as, salmon, Spanish mackerel, skipjack, sardine etc.

Characteristics of Gill Net Gear

Considering the process of catch by gill net, the gear should not be noticed by the fish. The fish which are enmeshed or entangled in the net should not be able to escape. So, material of netting cord, mesh size, hang-in ratio and tension are important factors need consideration.

Material of netting cord is very important for gill net fishing efficiency. Earlier soft natural fibre, e.g. hemp, rummy, cotton and silk are used. Flexibitity of the cord with high breaking strength is also required for gill net fishing operation. Nylon fibre is proved to be more effective in all respects, such as, flexibility, breaking strength, elongation, colour and durability than the natural fibres. At present, almost all gill nets for salmon use the nylon net composed of multi-filament or mono-filament, mainly because of its flexibility, high efficiency of catch and it does not decay in water.

The breaking strength of the gill nets should be as high as possible so as to prevent breakdown while fishing. The fish when gilled or entangled in the net exert some momentum in attacking the net, so they may escape from the net or break the net if the breaking strenght is not enough. It is desirable to select netting cord which has high breaking strength but low thickness.

Proper elongation of netting cord is necessary for gill net to catch fish. Normally

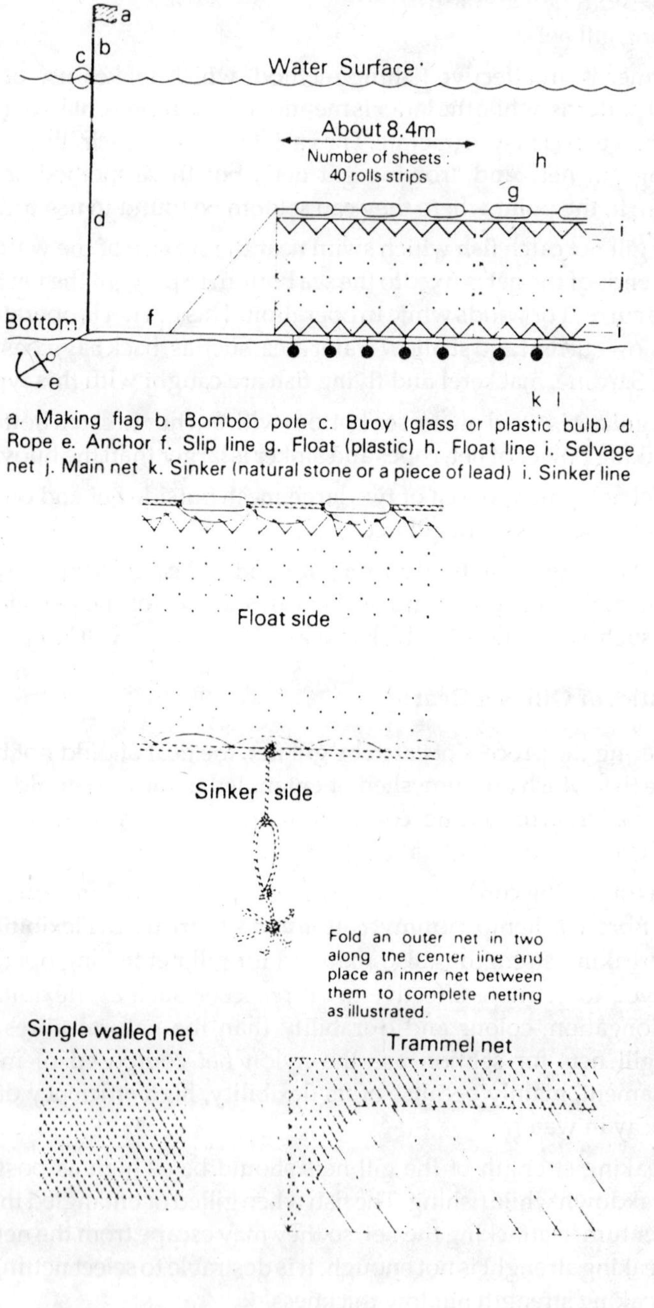

Fig. 77: Characteristics of gill net fising gear

the elongation ratio, i.e. the elongation at the breaking point divided by original length is about 25 to 30 per cent. The elongation ratio of various synthetic fibres are:

Kind of twine	*Elongation ratio (%)*
Nylon 210 denier	26-30
Polyvinyl alcohol (kuralon) 20's	25-30
Vinyliden chloride (Saran) 360d	20-25
Vinyliden chloride (Kurehalon) 360d	22-28
Vinyl chloride (Tevilon) 300d	18-23
Polyester (Tetoron) 210 d	14-18

The catch of salmon and trout in northern waters has been increased considerably by changing to a thinner, less visible twine in net construction. The elements involved in invisibility are colour and transparency. The twine used in the net should be such that it does not offer a contrast to the environment in which it will be used. For example, nets used for catching lobsters in reef areas are made of reddish twine; bluish twine is used for gill nets that catch flying fish in surface waters of high transparency; and dull grey or dark green twine is used for nets catching sardine or salmon in the dark middle-depth waters. Lately nets made of nylon monofilament are performing very well because of their high transparency.

Net's characteristics do not depend simply on the material from which it is made, but also on the structure of the twine (the thickness, no. of yarns, the type of twining and the type of knotting used). All these factors cause slight differences in the characteristics of the netting as a whole. It is clear that the catching capability of a net depends heavily on its elasticity, degree of elongation and return rate of elasticity. More elastic netting twine means that have a larger choice of mesh sizes for catching a given size of fish.

Trawler knotting is suitable for a large-mesh gill net, so that it can not slide off easily when too strong a pulling is given to either knotting leg, and is completely stabilized.

Considering the factors involved in the net structure, the only way to design a net that will perform as it should in actual fishing conditions is to base that design on knowledge that comes from years of actual experience. However, there are some principles that can be used in designing a net that will have a good chance of being successful.

The catchability of a net can be expressed in terms of such factors as mesh size and shape, rigidity of mesh leg and entangling function, but the truly decisive factors are the following three:

(a) Mesh specifications

(b) Hang-in-ratio

(c) Tension applied to the netting.

Mesh Specifications

Net functions simply to strain the fish out of a certain area of water in case of boat seize, purse seize, lift net and set net. As such, any fish that is larger in length or girth than the mesh size will be caught by the net. But gill netting calls for a unique performance on the part of the net.

Any fish which is too small for the mesh dimensions of gill net will be able to slip through the net and escape, while any fish that is too large on the other hand, will hit the net without getting enmeshed and be able to escape the way it came. For this reason the size of the mesh in gill net becomes a factor in selecting the size of fish that will be caught.

The gill nets in use today vary greatly in mesh size from 4.2 cm (for prawn, flying fish etc.) to 18 cm. (for shark, tuna, marlin etc.). But the common nets used to catch a variety of fish in shallow sea areas tend to fall into the 5 cm to 10 cm. range.

Hang-in Ratio

In order to ensure that the gill net will hang with a proper mesh pattern that will effectively receive the fish, the net is sewn to a rope that is shorter than the net length in a suitable ratio. This ratio at which the net is shortened, in other words, the hang-in ratio, is a very important factor in the gill net's operation.

The hang-in ratio is directly related to the body shape of the fish to be caught. Compared to the nets used to catch pelagic, fishes with a basic (spindle) body shape, nets used to catch slim profile bottom inhabiting fishes have a larger hang-in ratio. Furthermore in the case of nets for catching flatfish family and crustaceans, where it is easier to catch by means of entangling, rather than enmeshing, nets are usually made with a hang-in-ratio of greater than 50 per cent.

Tension applied to the Netting

When the gill net is suspended in the water, a floating force and a sinking force is acting on the net. In order to keep the net suspended in a given position, there must be a balance of floating force (created by the floats) and the sinking force (created

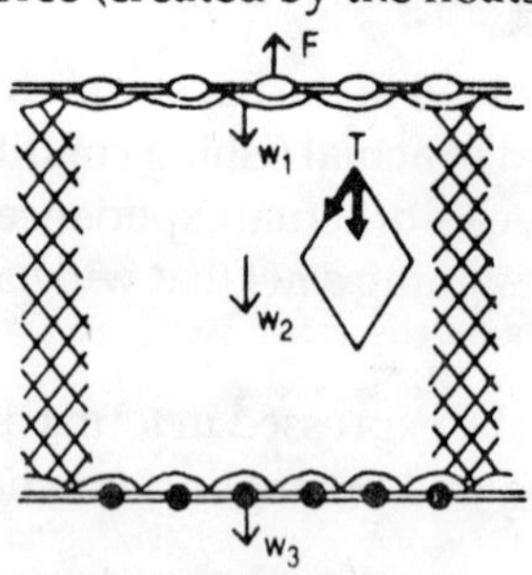

F: floatational force of the floats
w_1: weight of the float rope
w_2: weight of the netting
w_3: weight of sinkers and sinker rope
T: tension of the net cord

Fig. 77: Model of performance changes in gill net

by the sinkers, weight of the net and the ropes). In addition, a number of other forces, such as, the weight and pulling force of the fish caught in the net, the force of tide etc. must also be taken into account. The most important of all is the effect of the tension applied to the legs of the meshes by the floating and sinking forces. A gill net holds the fish which have become enmeshed by means of the tension exerted by the legs of the meshes. If the tension applied by the legs is insufficient, it becomes easier for the fishes to work themselves free of the net. Gill nets used to catch bottom fishes, such as flat fish and shark must be stretched with greater tension, while on the other hand, drift gill nets for Spanish mackerel or prawn gill nets which catch mainly by entangling, should be set with as little tension as possible, allowing a large hang-in ratio and plenty of slack in the net overall.

Gill net's catching effectiveness depends on the way the net hangs in the water. The desired hang is a straight vertical line or a loose curve along the sea floor, depending on the type of fish to be caught. These factors must be taken into account when designing the floating and sinking forces that will work on the net.

Design Particulars for Effective Gill Net Operation

The basic aim of gill netting is to get the fish to enmesh itself in the net. When the fish enters a mesh in the net head-first, the thread of the net must have enough elasticity to allow the fish to continue to move through the mesh to a point where the tension at the mesh threads equal the muscular strength and flexibility of the fish, after which the fish is held in place and caught. However, in the case of big fishlike marlin or tuna or fish with of body shapes, such as, flatfish and flounder or various crustaceans, which make it difficult for the body to enter a mesh eye, the main catching ability of the net is to make the fish "entangle" themselves in the net rather than enmeshing in one mesh.

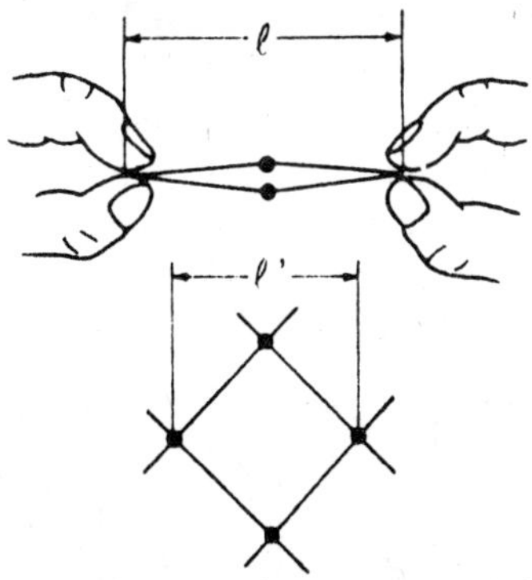

$$\text{Hanging ratio} = \frac{\ell - \ell'}{\ell}$$

Fig. 78: Hanging ratio of gill net

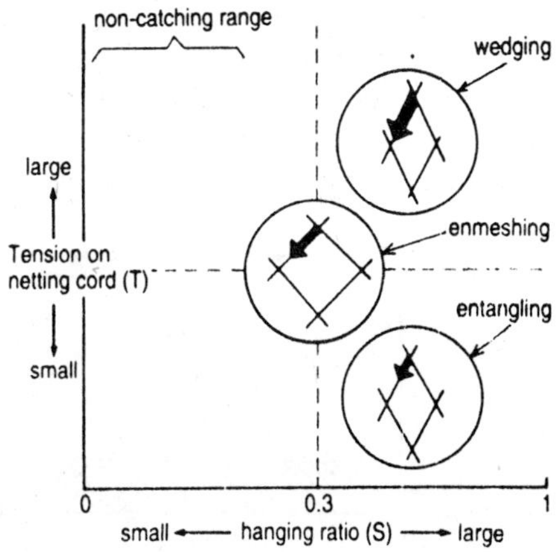

Fig. 79: Tension on gill net cord

The net is constructed from a single piece of netting, called the body of the net, to which a selvedge is sewn along the upper and tower edges. The selvedge portion is made from a piece of netting with heavier twine and smaller mesh size.

Materials that are necessary for a gill net, besides the body net material include, ropes, sewing thread, floats, sinkers, buoys, buoy line, anchors, and anchor ropes. All netting and ropes are now being made of synthetic fibres. Floats are usually made of foam plastics body and sinkers usually of lead. Synthetic fibres have proved themselves to be superior to the natural fibres with respect to light weight and durability. As regards floats and sinkers, similar performance can be achieved with glass bulbs, cork and wood and natural stones in case of sinkers.

Bottom Set Gill Nets

The net is set in open sea areas at a depth of 40-50 metres to catch tongue sole in areas with sandy bottom and muddy bottom areas to catch blue crabs. The net is also used primarily to catch sea bream that migrate between deep and shallow water areas in spring and winter seasons.

The net is set in a straight line at right angles to the flow of tidal current. After throwing in a buoy at the site where the setting will begin, the boat backs away at a speed of 2 or 3 knots, casting the net into the water from the port side of the boat. When dropping in an anchor, the anchor is first held at both ends, with the legs spread apart, and given a firm pull to stretch out the sinker rope before it is dropped in the water. The net setting opertation should take just over 10 minutes. For a normal gill net two anchors are used, one at each end of the sinker rope. In waters with strong tidal current, one anchor is attached at every 150 metres along the sinker rope.

During hauling the net, the rudder is first turned to port and set in that position. The main engine is used to run the generator, which powers the net hauler in the bow of the boat. When the float rope reaches the roller, the net is hauled in hand over hand by two men, one on the float rope and the other on the sinker rope. As it is hauled up the nets piled on the deck, folding it as it is piled. The anchors that are hauled in must be neatly lined up at one corner of the deck. During this process, the fish that have been caught are freed from the net. In the case of three-man operation, the third man stands in the middle of the deck and frees fish from the net as they appear. The hauling of one unit km. net takes between 60 to 90 minutes.

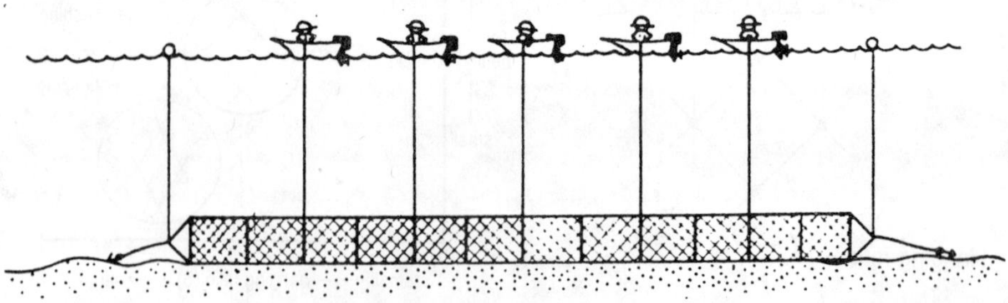

Fig. 80: Bottom set gill net

Drift Gill Net

The fishing grounds for Spanish mackerel drift gill net are found both in main

bottom. After about half an hour of drifting, the net is hauled in and the catch is shaken out of the bag net. This process is repeated several times during the night.

The configuration of the net during operation can be classified into four types:

Condition A — This is the ideal net shape, with a large mouth opening. As the relative speed of the net is slightly slower than the tidal current, the leading edge of the bagnet stretches backward for some distance from the sinker rope, resulting a smooth upward curve of the bagnet. This makes it easy for the prawn to enter the net and hard for them to escape.

Condition B — Often this is a condition of the net during operation. When the relative speed of the net drift is almost identical to the speed of the tidal current, the bagnet inflates upward from the sinker rope. Since there is no backward stretch of the net from the sinker rope, this condition is inferior to condition A.

Condition C — In this case, because the bagnet is too soft, it is flattened out by the tidal force in the forward direction. The result is a depressed mouth opening.

Condition D — Both the body net and bagnet are pushed flat against the sea bottom, with part of the bagnet turning the inside out. If this situation continues for some time, the bagnet becomes entangled in the sinker rope and the bagnet's function is completely lost.

Salient features of gill nets

Gill net type	*Mesh size (cm.)*	*Depth of net (meshes)*	*Hang-in ratio (%)*	*Float line buoyancy (g.)*	*Per metre sinking force (g.)*
Floating gill net					
Sardine	3.6	100	30	1120	200
Shark	30.3	35	41	128	30
Drift gill net					
Herring	5.5-7.3	60-200	35-50	425	45
Young yellowtail	9.0	100	45	55	35
Tuna	18.2-36.4	20-70	35-50	285-435	112-178
Bottom gill net					
Herring	5.2-6.7	40-60	50	300	1680
Shark	14.5-24.2	14-40	30-50	270-380	725-1020
Sea bream	16.7	12	53	10	15
Flat fish	14.2	15	48-68	2.5-15	10.35
Lobster	10.6	35	33	20-30	32-45

Gill Netters

Gill nets may be bottom set, surface set, anchored or drifting but these variations do not affect the vessel layout much. As with lines the nets are set over the side or stern, and hauled in over the bow. The fish are picked out or shaken out as the nets come on board. A gill net may measure anything from one metre to ten metres in depth. Though single net may seldom be more than 100 metres in length, yet the combined length of a fleet of nets may stretch for several kilometres.

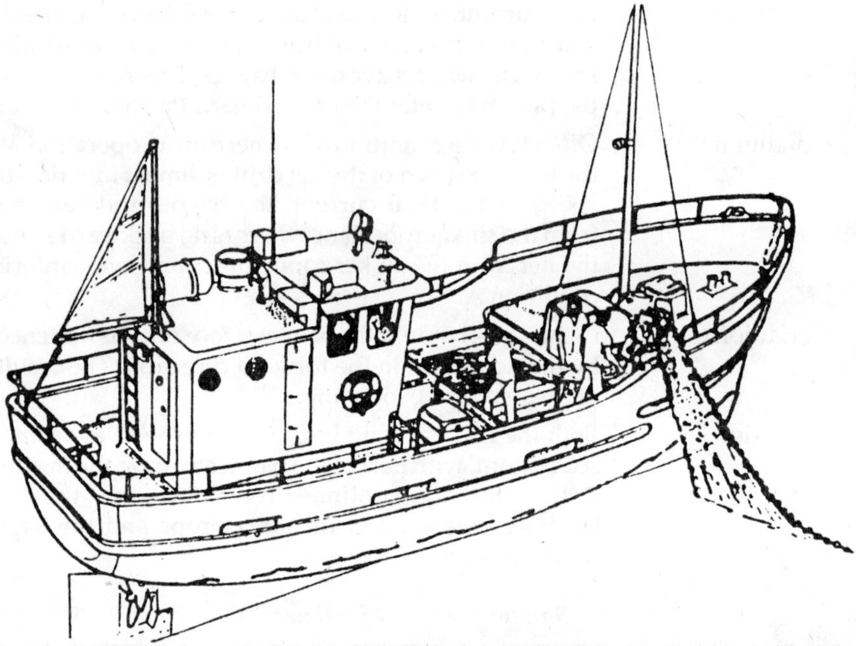

Fig. 84: Gill netter

The main problem with all gill nets is that of preventing entanglement or fouling. Firstly there should be nothing on the vessel to snag on the net as it is set or hauled. The rail or bulwarks should be clear of even the smallest bolt or nail head that might catch on and tear the gear. Inside the boat all cleats, hatches, fittings, corners and protrusions near where the nets must come should be tucked out of the way, smoothed over or covered with heavy plastic, rubber or canvas.

A lot depend on the types of net used and the size of its meshes. The float should be shaped so they will not slip through the meshes. The weights should be of a design that will not foul either. Ring weights may be used, or even leaded rope, to avoid fouling. Some gill netters use a pipe or bar or shooting fork on to which the ring weights are strung so they may be set in order like a series of purse net rings.

Most gill nets used to be hauled by hand. The first powered hauling of gill nets was the use of a steam capstan in the great herring drift net fishery of the North Sea. The capstan was used to haul in the heavy rope attached to the foot of each net by short tie lines. This made it much easier for the crew to pull in the netting. A long rail roller with a one-way racket was also fitted to these vessels.

Modern gill net haulers have powered sheaves somewhere between a line hauler and a power block. Net drums are also used for gill nets and their applications require hauling in over the stern. Care must be taken not to damage the fish when pulling nets in by powered reel or hauler.

Specification of gill net per unit

Net type & fish aimed		*Mesh size (cm.)*	*No. of mesh*	*Length of net (m.)*	*Float length (mm.)*	*Sinker length (mm.)*
Drift gill net						
Sardine		4.5	250	75.0	50.0	60.0
Mackerel		7.3	500	60.6	40.9	42.4
Bottom gill net						
Horse mackerel		7.5	100	50.5	36.4	33.3
Flat fish		16.5	50	195.0	117.0	117.0
Shrimp		6.1	22	75.8	29.0	38.0
Blue crab		12.1	17	75.8	35.0	35.0
Lobster		6.1	17	25.0	8.3	8.4
Trammel net						
Flat fish	out	33.3	4	100.0	71.2	80.3
	In	7.0	25	151.5		
Shrimp	out	24.2	6	77.3	34.9	37
	In	6.1	30	75.8		
Lobster	out	30.0	3.5	50.0	36.0	38.3
	In	9.0	17	72.0		
Bream	out	37.8	3	91.0	27.0	27.0
	In	9.0	20	70.5		

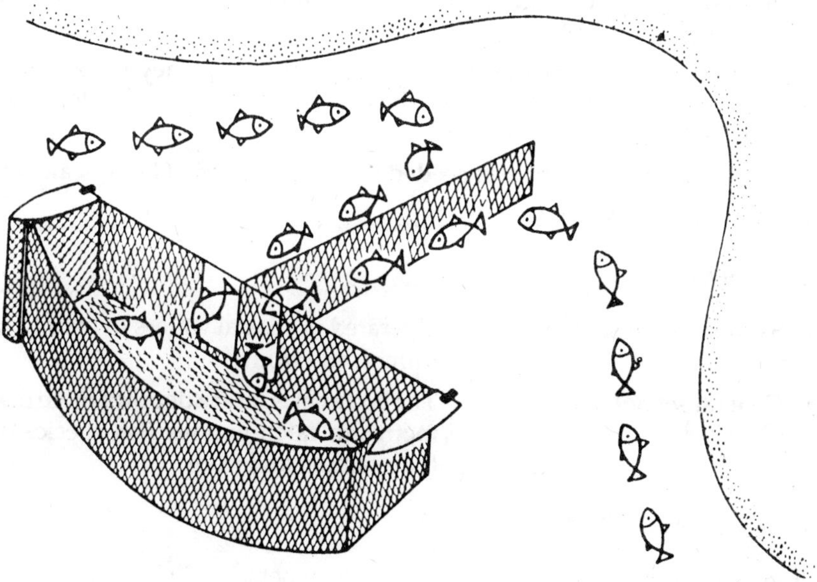

Fig. 85: Set net in operation

9

SET NET

In set net fishery, the net gear is kept fixed in some specific areas for a long tim so as to get fish or crustaceans regularly every day when they come into the bagne unconsciously. This type of fishery is, therefore, called "wait and see" fishery i contrast to such aggressive types of fisheries, such as, trawl net and boat seine whic are operated to aim at the moving shoots of fish by net. Set net fishery is subjecte to the influence of various natural factors including sea-bottom nature, stream con ditions and seasonal changes in migratory fish species. The advantages of set ne fishery are:

The net is set in such a manner that it can continuously catch the fish for a lon time until it is pulled up, that is, fish schools swimming with the stream are lure into the lag net by means of the tactfully set-leader line. In this way the fishin operation is continued day and night. The net is dragged up once a day, usuall in the morning (twice a day at some fishing grounds, in the morning and in th evening) and the operation can be completed within two hours. The number o fishermen needed to depend on the size of a net and in case of small-scale set ne operation it does not exceed five or six.

Many different species of fish are lured into the bagnet and they are caught aliv Drastic fluctuations in the catch is obtained due to the seasonal behaviours o migratory fish species.

Large amount of fixed capital is needed for the operation of large-scale set ne

Set net fishery is divided into two types:

Large-scale Set Net

(i) Salmon/trout set net — Operated in coastal waters to mainly catc salmon and trout.

(ii) General set net — The body net is set at a depth of more than 2 metres to catch various fish species othe than salmon and trout.

Small-scale Set Net

All set net operations other than large scale set nets.

Basic Elements in Set Net Fishery

Basic elements involved in set net fishery operation are: (a) fishing grounds, (b) fishing gear, and (c) management system.

Choice of Fishing Grounds

In order to locate a suitable fishing ground for set net, it is necessary to study the types of fish species present and their seasonal migrating habits in the coastal waters. The success or failure of set net operation is determined by the position and angle of net setting in the "fish route", i.e. the areas or routes where schools of fish pass with high frequency.

The important factors that determine the configuration of the "fish routes" are: (1) the characteristics of coastal isobath, (2) the configurations of natural reefs in the area, (3) the sea bottom topography and its interrelation to the currents and (4) the consistency of the sea bottom (rocky, sandy etc). From a good knowledge of these inter-relating factors it is possible to estimate the types of fish, size of schools and seasonal changes that will occur in a given fish route.

The Coastal Isobath

In the case of large-scale set net fisheries catching mainly migrating species, such as, yellowtail, horse mackerel, mackerel, sardine and anchovy, the areas chosen as fishing grounds are the steep coast areas near the mouth of deep bay, where a branch of the ocean current that has entered the bay and circled it is about to leave the bay again. Fish species that are migrating with the circulating currents along a coast form fish routes and because each species migrates at a given depth, when they come to a place on the coast where the isobath lines are close together it is believed that the schools of fish are forced together and thus become concentrated in a small area.

The Existence and Configurations of Natural Reefs

The natural reefs of the coastal waters though beneficial in that they lure migrating fishes, yet they also function as substantial obstacles. In this situation leader nets all point toward a reef, or are positioned with their body nets next to a reef or between two reefs, so that the location of reefs is directly related to the way the nets are positioned.

Effects of the Tides on Fish Behaviour

There are a variety of traditional fishing methods that take advantage of the movement of fish and shell fish with the tides in shallow sea areas, which are changed into an expansive tideland at high tide.

Areas in which inland marshes, lakes or rivers are connected to the shallow sea by estuaries provide excellent nursery grounds for the fry and fingerlings of various ocean fish species, due to the abundance of nutrients that these inland water bodies are sending into the sea water, and also due to the constant tidal exchange with the open sea. Therefore, these areas make ideal fishing grounds for small scale set net fishery.

Physical Characteristics of the Sea

In addition to the above-mentioned factors there are a number of physical properties of the sea which is to be considered.

Sea depth	—	With today's technology, the deepest level at which a set net can be set is about 100 fathoms (183 metres). The deeper one goes, the stronger the forces of water stream become, thus tending to alter the shape of the net. Therefore, the net must be larger and heavier, which means that net hauling capability becomes another factor limiting the depth at which a net can be set.
Set current	—	Most nets need to be set in areas with relatively mild water current. If the current power is too strong, it tends to distort the shape of the net, thus lowering its catching efficiency; and also tends to cause various difficulties in the hauling operation. The maximum current power under which net hauling can be successfully performed is considered to be 0.4 knot/hour to 0.6 knot/hour for an under water set net.
Wind speed	—	A set net fishing ground must be an area with mild wind and wave conditions. Areas with strong winds invariably have rough wave conditions which hinder the net hauling operation and increase the rate of breakdown of the net and other fishing gear.
Sea bottom	—	Rocky sea bottom areas are not suitable for set net operations, due to the possibility of net destruction. The best sea bottoms for set net are mud, sand, pebble and shell, or a combination of these.

Selection of Fishing Gear

The history of the development of set net technology centres around two conflicting objectives: (i) to make a net, which is easy for the fish to enter, and (ii) to frame a net from which it is difficult for the fish to escape once they have entered. For basic net design then, for each region, improvements and changes are based on the habits of the fish species, which is decided will be most profitable to catch considering the make-up of the fishing grounds and fish resources of the particular area.

A set net is made up basically of four parts:

1. Leader net — The function of this part of the net is to obstruct the course of migration of schools of fish and to stimulate the fish and lead them in the direction of the body net.
2. Body net — This part of the net serves to stop the schools of fish and hold them. It includes a space called the play ground.
3. Bag net — This part of the net gathers fish and prevents them from escaping.
4. Funnel net — This part supplements the function of the bag net. It is a trap, which lures the fish in and then cuts off their route of escape.

Based on their basic construction, set nets can be divided into five groups as given below:

Group A — This is a prototype, large-scale set net, which is seldom used today. Having the bag net included as part of the body net, when a school of fish was found to have entered, the mouth of the body net would be closed and the catching operation would proceed.

Group B — This is the main net configuration used today in both large-scale and small-scale set net operations. It is an improvement on the net of group A.

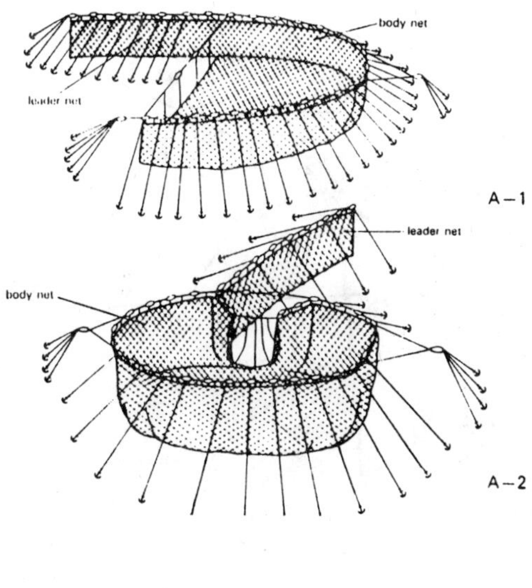

Fig. 86: Large scale set net

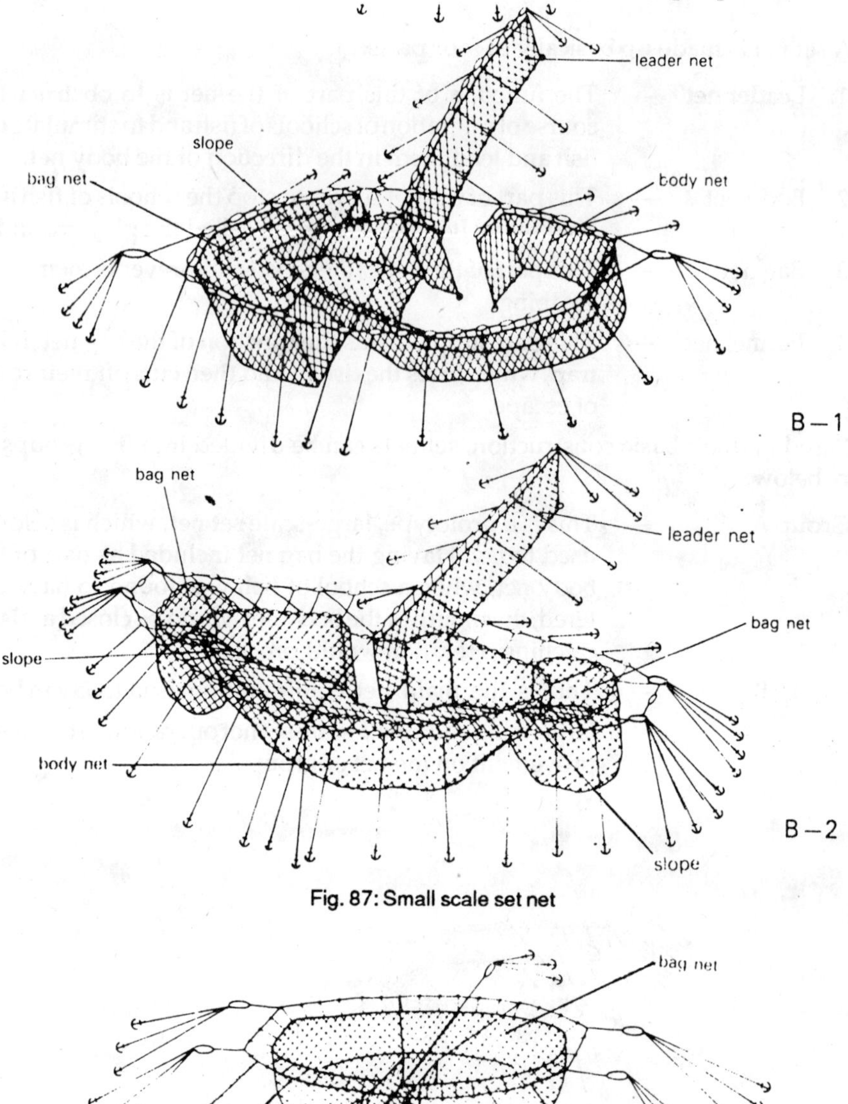

Fig. 87: Small scale set net

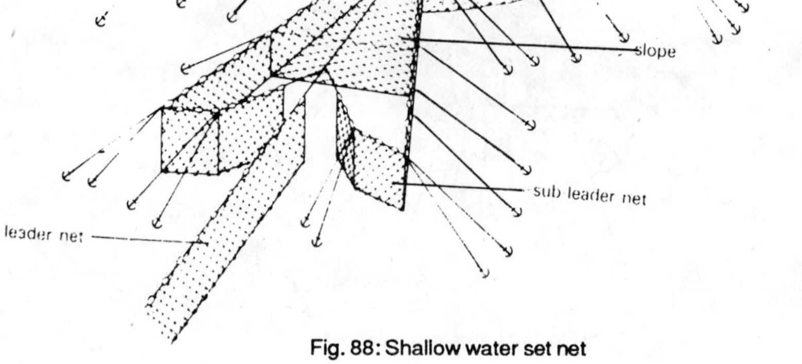

Fig. 88: Shallow water set net

Group C — Compared to the group B nets, this net is better suited for shallow water areas or areas where the fish migrations are spread out thinly over a large area.

Group D — This is a small-scale set net that was developed in the 17th and 18th centuries in the shallow water fishing grounds. They usually feature one or two funnel nets fitted in each bag net.

Group E — This includes most primitive types of set net configurations. In some cases, the basic layout of the net is not clear, because they function by making use of some unique natural characteristic of a given fishing ground. For example, in E3 a tide pool functions as the bag net, and in E4 a current in a shallow sea area serves the function of the leader net.

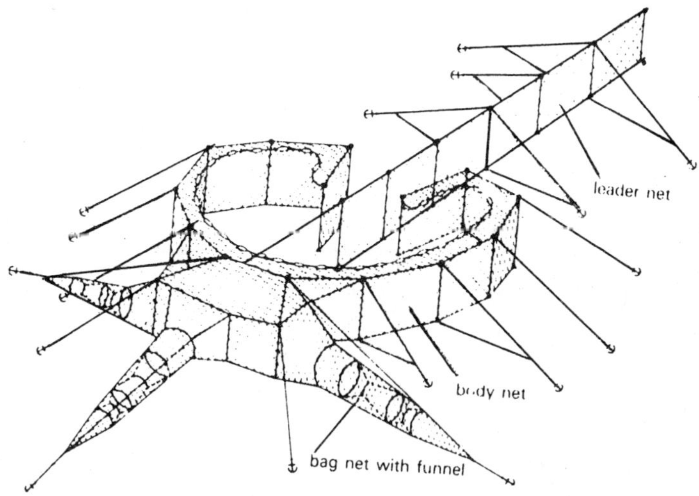

Fig. 89: Set net fitted with bag net and funnel

Once a basic type of set net has been decided on, the net is designed with consideration given to the place in which the leader net will be set, the proper size and shape for the mouth of the net, the size of the "playground", the proper angle for the slope, the dimensions of the funnel net and the volume of the bag net. Also depending on the type of fish one is trying to catch, study must be made of the way the fish reacts to the net in order to determine the proper twine thickness, colour (brightness) and mesh size to be used in the net construction.

Choice of Operation

Selection of suitable fishing grounds, predicting annual catch and planning reliable programmes of income and expenditue are the prerequisite for choosing set net fishery, in view of large capital investments required for large-scale set net, boat

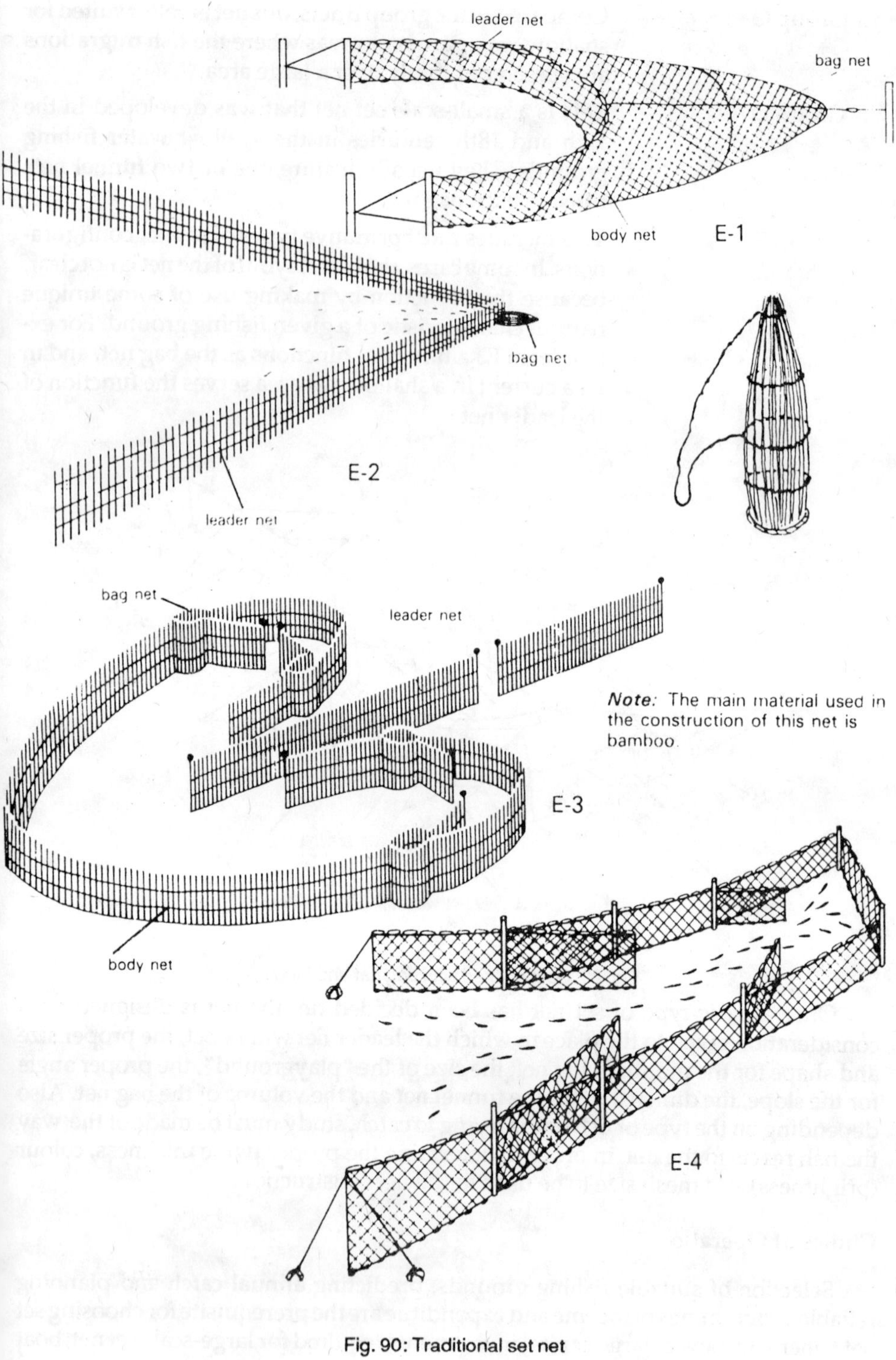

Fig. 90: Traditional set net

and labours and also the increased danger of having the nets washed away by strong current or rough seas.

While planning introduction of set net fishery, the first step would be to initiate small-scale set net operations. Basing on the operational experiences gained, one can investigate the feseability of introducing large-scale set net.

The individual set net fisherman must work to imporve his productivity. A fisherman who neglects this responsibility will not only see a deterioration in his own productivity, but is also likely to be critcised by neighbouring fishermen for the abuse of his fishing rights.

Recently the set net fishing grounds in coastal waters are being destroyed by a number of phenomena involved in the industrialization, such as, the reclamation and urbanization of shallow sea areas and the resulting decrease in water quality and the direct dumping of waters from factories into coastal waters.

Modernization of Set Net Fishery

The underwater set net and bottom set net have come to be used in many fishing grounds due to the adoption of power driven machinery and equipment since late sixties. The aims of using these nets are:

(a) to increase the amount of catch by preventing fish from escaping out of the bag as effectively as possible.

(b) to increase the species of fish caught by catching demersal fish as well.

(c) to develop offshore fishing grounds by adopting these seaworthy set net methods.

The two-mouth bag net is considered most suitable for the shoal extending to a great distance, where fish species are spread thinly over a large area, while the one-mouth bag net is mostly used for deep coastal waters with the aim of making the schools of fish concentrated in a small area along their routes.

Improvement of fishing gear is made by:

(a) Improvement of net material (from jute yarn to synthetic fibre).

(b) Development of double bag net.

(c) Adoption of large-size buoy (made of steel in place of bamboo or wood).

(d) Improvement of small-size buoy (from glass balls to plastic material).

(e) Improvement of setting system (synthetic fibre as sand bag material, instead of natural fibre and modified metal anchor or concrete block holding, instead of sand bag system).

(f) Improvement of lead line (synthetic fibre rope using lead as its core).

Machinery and tools (fishing) have been improved by improving the net hauler (manual winch-ring type mechanical net hauler-catch hauler); adoption of high performance fish finder and labour saving method of net setting.

Application of dye stuff on fishing net to keep off sea weed in another improvement made in set net fishery now-a-days.

In small scale operation, two bag nets were fitted on opposite sides in an attempt to increase catch effectiveness. Because of this, two-body net entrances were created on either side of the leader net, where before there had been only one.

The leader net length was increased from 105 metres to 130 metres to increase the number of fish schools intercepted by the leader net.

The slope was also improved. The angle of slope was reduced to a gentler angle in order to decrease the fish's resistance to entering, and thus increase catch efficiency. In addition, a ceiling net was strung over the top part of the slope to prevent fish from escaping.

Four pulleys were fitted on the ropes supporting the bag net. Two ropes are strung through these pulleys and under the lower part of the bag net. When these tow ropes are pulled by a power winch, they performed the same net raising function that had previously been performed by the crew raising the net, hand over hand in the direction of the upper part of the slope.

The width of the body net mouth is made larger. These improvements (addition of pulleys and power winches) has simplified the hauling operation, making it possible for four fishermen to perform the job, which previously required 5 or 6 men. With the help of these power winches adjustment and maintenance of the shape of the net can be performed more accurately. With the new net configuration a large number of demersal fish begin to be caught along with main catch of surface dwelling fish, meaning larger over-all catches by each net.

Small-scale Set Net in Lagoon

Lake Hamana is a salt water lake connected to the Ocean by a 200 metre wide channel in Japan. It has a total area of about 70 square kilometres. The average depth throughout the lake is 4 metres. Having a rate of exchange with the Ocean water at the channel mouth of 4200×10^4 cubic metre per day, the lake has a very high salinity of 15–16 ppt throughout. Although there is a difference of about 80 cm. between high and low tide at the mouth of the channel, there is almost no change in the water level due to tides, a little distance inside from the lake mouth.

Fisheries of the lake are supported by two groups of marine life: (1) species that live in low salinity water area; and (2) species that circulate between the two environments of the open sea and inland waters in their life cycle.

The fishing grounds are sandy bottom areas with a depth of less than 4 metres and the fishing operations are concentrated in the southern half of the lake with the help of small-scale set nets.

Fishing Gear

The small scale set net used in the lake consists of (1) a leader net, (2) a body net hung in a triangular configuration, and (3) three bag nets. The leader net is laid out parallel to the direction of the tidal flow, with the mouth of the body net usually being set on the down-tide side. However, different types of settings are used depending on the season.

In spring and summer when large numbers of fish come into the lake from the Ocean, moving around the lake in a daily pattern that follows the ebb and flow of the tide, a pair of nets will be set facing each other to take advantage of the two tidal movements each day. On the other hand, in autumn, a single net is used to take advantage of only the outgoing tide. The purpose of this setting is to catch the prawn that have been raised in the lake waters but begin to migrate toward the ocean in autumn.

Fishing Boat

The boats used are 1.5 tonne F.R.P fishing boats, equipped with a 12 h.p. diesel engine.

Fishing Operation

With the exception from Jan. 16th to March 24th, when fry and fingerlings are migrating in the lake, fishing is carried out throughout the year. The bag nets are emptied of fish once a day in the morning. One or two men are required for this operation. In the summer, to prevent seaweed from growing on the nets, the nets are taken out of the water in the morning and hung out on poles during the day and then put back in the water in the evening.

Bag Nets

The mesh size of the net used in bag net varies from 28 to 43 mm. depending on the type of fish being caught. The larger the mesh size, the brighter the interior will be, making it easier for large fish, such as, sea bass, sea bream, flat fish and flounder

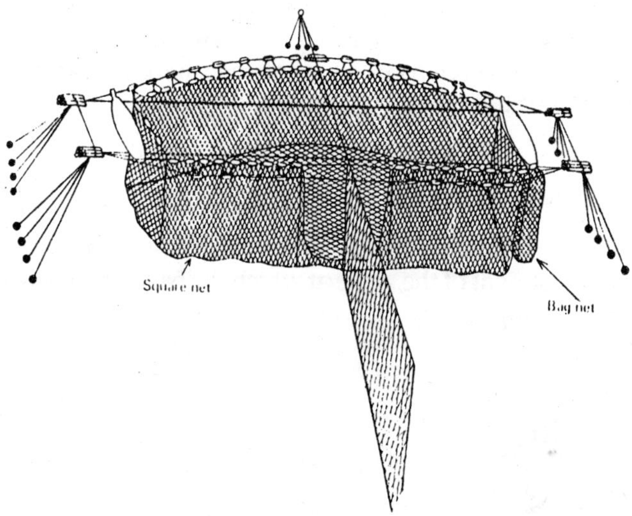

Fig. 91: Set net with bag

to enter the net. However, if the meshes are too large, fish will tend to get enmeshed in the net, making the hauling operation extremely difficult. When fully assembled, the over-all length of the bag net is 8 metres.

The funnel net is attached to the bag net just three mesh eyes in front of the position of the hoop of vinyl pipe. If it is directly attached to the hoop, it will become too rigid, thereby increasing the incidence of injury or death to fish struggling to escape the net.

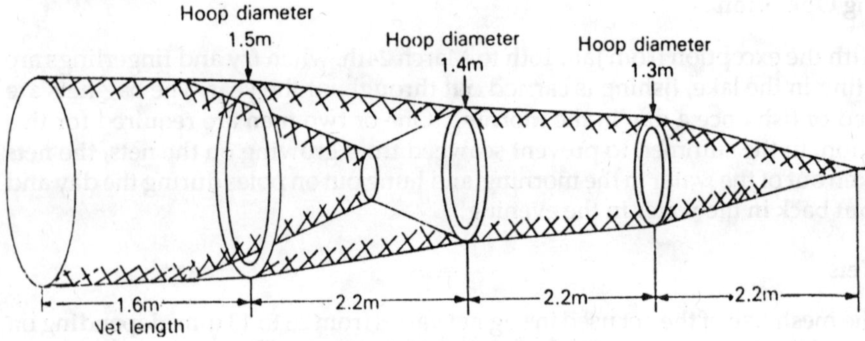

Fig. 92: Bag net specification

Fitting of Bag Net to the Body Net

Staring at centre point four-tenths of the way up the net from the bottom, a vertical and horizontal line are cut in the net in a "+" pattern. Four triangular pieces of net are prepared. These are then fitted into the cut sections of the body net and sewn in places. The ends of the triangles are then sewn to the periphery of the bag net. This creates a smooth curved connection between the bag net and body net when the bag net is pulled out tight, reducing the fish's reluctancy to entering the net to a minimum.

Hanging the Bag Net

A pulley is used to facilitate adjustment of the tension on the bag net, as well as to allow for easy adjustment of the angle at which the bag net is hung according to the habits of the fish being caught. However, in any case, the bag net must always be kept off the sea bottom.

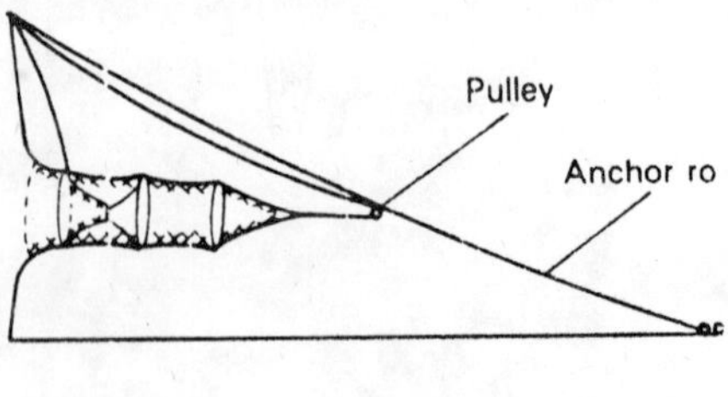

Fig. 93: Hanging bag net

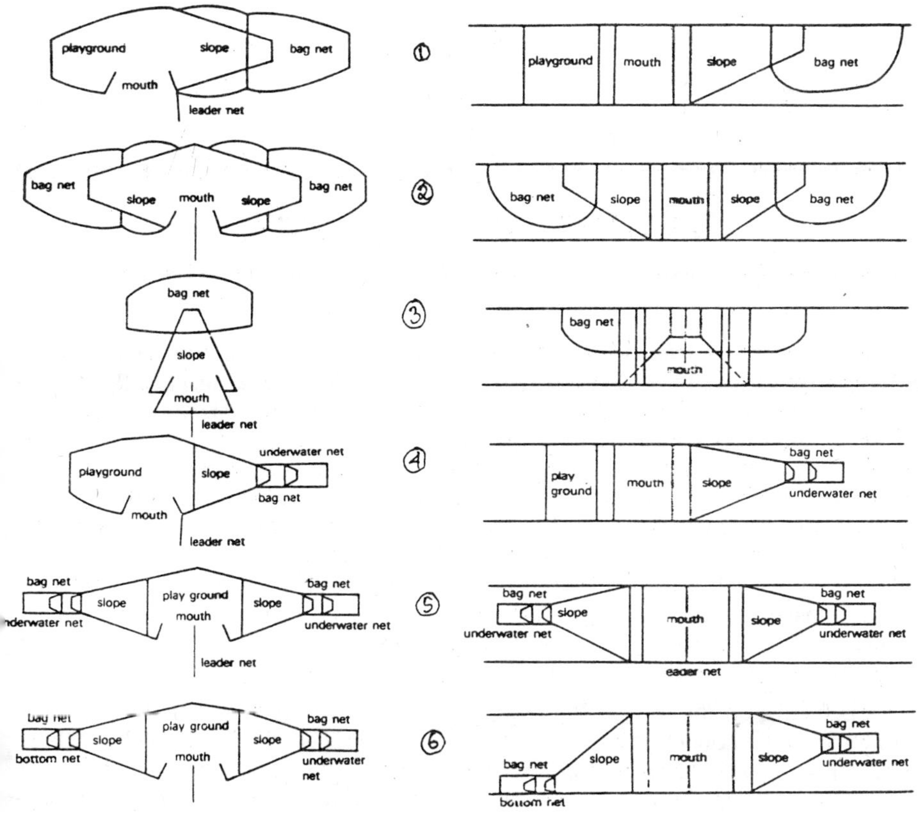

Fig. 94: Two dimensional sketch Three-dimensional sketch

1. One-end bag net; 2. Both end bag net; 3. Choko-ami; 4. One end under water; 5. Both end under water and 6. One end under water and the other end bottom set.

10

BARRICADE FISHING

Jano fishing is a traditional way of enclosing fish in shallow water areas of Chilka lake in Orissa, India. In Jano fishing, a large shallow water area of the lagoon is enclosed by bamboo screens. After the monsoon, the water level increases and lots of low-lying becomes submerged under water for 3 to 6 months. These areas with the influence of the sunlight and high temperatures become highly productive. In September-October when the salinity has increased a little, many fish species migrate from the Bay of Bengal to feed in these shallow water-areas. The fishermen watch these migration of fish and when the fish has come to the shallow areas, they enclose it with bamboo screen. The area of the jano varies between 13 to 813 hectares. For the closing of jano 70–100 fishermen is needed and the closing is done very quickly. In deeper water, 30–40 boats are required to fix the bamboo-screens. The closing of such jano is done within few minutes.

Depending on the size of the jano, it has one or more catching-areas, called gohoras. This is a construction of two parallel bamboo—or net screens about three metres apart. On the inner screen there is a small opening. During high tide fish swim against the current and finds its way out through the opening. Once inside the gohora, they hardly find their way back to the jano. The gohora is then emptied with the help of stick-nets, or bamboo mats. The gohora is usually emptied only once a day, but if there is a lot of fish, fish is caught twice or thrice a day. When very small size fish is trapped,they are thrown back into the jano to grow a little bit more. The

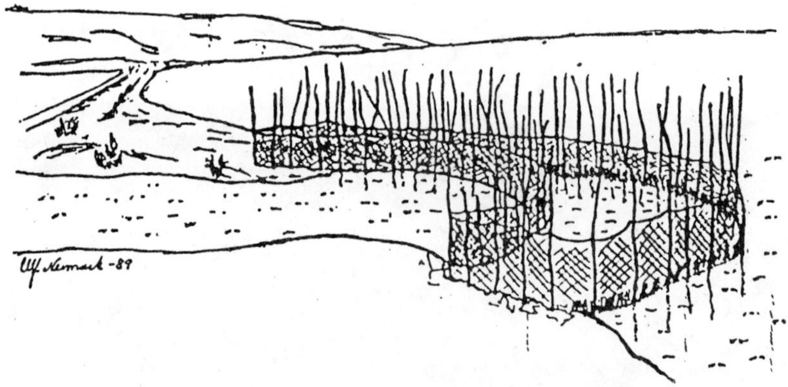

Fig. 95: Jano fishing

catching activity is done by the fishermen staying at the site of the jano. Depending on the size there can be up to 12 persons. They guard, catch and repair when needed.

The jano is operated during the winter, from October/November to January. The jano fishing is by its nature a community fishing.

The jano itself is not disturbed during the main season. All the catches takes place in the gohora and sometimes also in a few prawn traps inside the jano, close to the gohora. At the end of the season, the fishing activities also move inside the jano. It is then emptied by traditional types of net fishing. After the season, the bamboo-mat enclosures are removed. As the season nears its end, most janos have again become dryland.

The dominant species, caught in the janos are *Mugil cephalus, Liza macrolepis, Lates calcarifer, Gerres setifer,* and *Mystus gulio.* In some 80 per cent of the catch could be of one species. Production from a normal jano is said to be about 124 kg/ha/season.

Most of the equipment for jano fishing has traditionally been made out of by-products from plants; wood, fibres, sap, leaves etc. To close a jano, screens, sticks and boats are needed. The sticks are made from bamboo (*Bambusa arundinaceae*) or casuarina at a length of about three metres and pointed in one end. The screens are made of split bamboo and vary in size, length from 5 to 10 metres and 1–2 metres height. For tying the bamboo-splints together, creeper plants are used.

The jano fishermen usually use the medium size (12–13 m.) non-motorised country boats, built by planks from sal-wood (*Shorea robusta*). The wood is very heavy and weighs 22 to 27 kg per cft.

The bamboo mat enclosures used in enclosing janos are of:

(i) 8 mm. mesh, 2m, high and 10 m. long

(ii) 5 mm. mesh, 1.5 m. high and 5 m. long.

(iii) 5 mm. mesh, 2.2–2.5 m. high and 10 m. long.

Trap and Pot Fishing

Octopuses, belonging to the order Octopoda are widely distributed in the seas from the frigid to the torrid zones, and also vertically from the coastal shallow waters to the deep sea of more than 1000 metres. Two hundred species of this order are distributed throughout the world, comprising of surface and bottom dwellers. Among these varieties of Octopuses, the benthic species (*Octopus vulgaris, Paroctopus dolfleini dolfleini, Octopus ocellatus*) with well-developed muscles are used for food.

Octopus vulgaris

Most popular octopus, having the highest commodity value. This species is widely distributed in the Pacific and Atlantic oceans, the North Sea, the Mediterranean Sea, the Red Sea and the Indian Ocean, i.e. in many seas from the sub-tropical to the temperate zones. The maximum size is 60-70 cm. in body length.

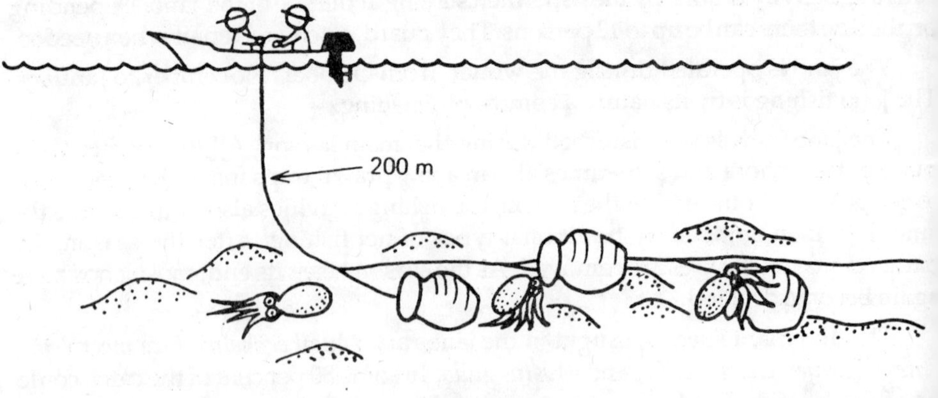

Fig. 96: Pot fishing

Fig. 97: Octopus pot

Paroctopus dofleini dofleini

This species lives in cold waters and in Japan this species is captured abur dantly. This species is characteristically large and individuals more than 3 metres i length are landed on rare occasions. It is said that this species, whose muscles contai much water is not so delicious compared to *Octopus vulgaris*, but it is one of tl important aquatic resources next to *O. vulgaris*.

Octopus ocellatus

This is a small species of about 10–15 cm. in body length and is abundant in tl inland sea. During spawning season (winter), the trunk becomes filled with egg

The spawning season of *O. vulgaris* is very long extending from the ear summer when the water temperature begins to rise until autumn. Hatched juvenil first spend a planktonic phase, and about one month after hatching they sink the bottom to start their benthic phases, when they begin to show negati phototropism. During the day time they hide in the spaces between or under rocks or other objects, and during the night they move about seeking food. They feed voraciously on crabs, shrimps, shellfishes, nereides and fishes, which they can catch at random. Octopuses live only in the sea with a salinity of 18 ppt. In sea water with 15 ppt. salinity the octopus shows abnormal behaviour and in salinity less than 15 ppt they die. The lowest water temperature in which *O. Vulgaris* can live is 6–7°C, and the optimum temperature is 22–25°C.

Octopus inhabits both rocky shores and sandy mud zones. Except during reproductive season, Octopuses do not form groups, and each individual has his own territory.

The growth rate differs with the environment, but generally the Octopus reaches a body weight of over one kg. at one year of age and becomes sexually mature. From the summer of the year after hatching the octopus begins to grow quickly and begins to be caught.

There are two fishing seasons in a year for octopuses. Major fishing seasons are from April to August (Summer) and from October to February (Winter). This has a relationship with the migration of Octopuses between deep and shallow waters. Octopuses prefer warm waters. So, when the water temperature drops in winter, they move to deeper waters, and when summer comes they come back near the coast from off-shore seeking warm water.

The migraton of *O. vulgaris* depend on the sea conditions or sub-marine topograpy. A change in salinity of the sea water induces the Octopus to migrate. In the inland sea, when the salinity of the water drops due to the increase in the amount of inflowing fresh-water from rivers in the rainy season in summer, many individuals of *O. vulgaris* escape to the offshore waters of high salinity causing a poor catch in the coastal fishing grounds.

Fishing Gear and Methods

The method of fishing take the advantage of the fact that the Octopus likes to enter in holes. An unglazed pot, a wooden or plastic box, or a concrete block with a lid is used as a trap. Shells are used to catch small *O. ocellatus*. These traps are tied with a branch line of about 30 cm. long; and these branch lines are further tied to a main line, and are laid on the sea bottom.

The size of pot or box differs according to the size of octopus being caught. Moreover, the total length of the main line and the distance between the pots also differ with the environment of the fishing ground, but generally the total length of one km. and traps are attached at intervals of about 10 m.

The fishing gear is laid on the sandy mud bottom or on the peripheral zone of reefs at a depth of 10–40 m. After one to three days, traps are pulled up by hauling the mainline hand over hand, and after taking out the Octopuses from the traps, the emptied traps are set again in the sea. This process is repeated throughout the fishing season.

This fishing gear is not suitable for use under conditions where the traps will be constantly rolled by strong tidal currents or an irregular surface on the bottom.

In the coastal waters extending about 450 km from Fremantle to Geraldton in Western Australia, 800 lobster basket fishing boats are in operation. This lobster fishing has been troubled with predation by Octopuses for several decades.

To catch Octopuses trap fishing has been tried, since trap fishing catches Octopuses alive, and traps can be hauled in at one's convenience. One man fishing is possible by using a line hauler. Various traps, such as, a pot, a piece of vinyl pipe, and a used tire were tried out, and the fishing gear made of a piece of vinyl pipe obtained the best results.

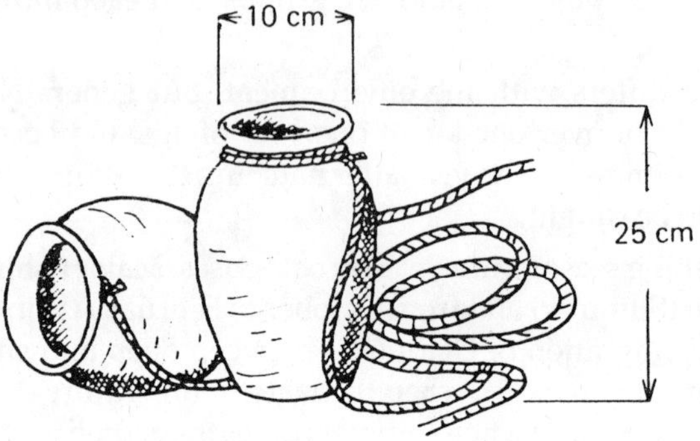

Fig. 98: Specification of octopus pot

Octopus pot fishing and Octopus box fishing are best carried out by fishing households using fishing boats of 3–5 tonnes. The reason can be explained by the characteristics of the resources and fish methods indicated below:

1. Because resources (Octopuses) are benthic, introduction of mass catch fishing methods by large fishing boats causes the rapid exhaustion of resources.
2. Octopus fishing is a passive or waiting type fishing method, and it can be carried out on a flexible schedule worked around the schedule of other fishing activities, and thus suitable for the family management.
3. Since the line hauling work has been mechanised, at present, the fishing operations can be conducted easily by labour power of one family. A fishing boat of 3–5 tonne can lay 500–1000 pots (or boxes) in the fishing ground in one trip. This number corresponds to the maximum daily labour power of a family.

Capturing Adult Eel

Adult eels inhabiting either rivers or lakes, are caught in autumn as they start to descend the rivers.

There are in total 19 species of eel in the world. Most of the eels are distributed throughout the tropical zone; however six species inhabit the temperate zone and two species both tropical and temperate zones. Two species of eel are distributed in the Atlantic ocean, one type inhabiting the side of European continent and the other the side of American Continent. As for the other species, 12 species are distributed in the Pacific Ocean, three in the Indian Ocean and two in both the Pacific and the Indian Oceans.

Eels are born in the sea and grow up in the fresh waters of rivers, lakes and ponds. Eels spawn in the middle layer zone of the Ocean where the temperature and salinity are high. Within 2 to 10 days after spawning, the eggs are hatched. The larvae, which are transparent and leaf-shaped, are called leptocephali. Leptocephalis, while moving up and down during day and night will spread out across the Ocean riding on the surface of the current and grow up as they migrate. In case of *Anguilla japonica,* the larvae will approach the coast about one year after they hatched (3 years for European eels) transform themselves into elver (fry) near the sea bottom, and soon after transformation, they enter the rivers. The water temperature strongly influence the migration to the rivers; and the elvers start to ascend rivers choosing the period when the difference of the water temperature between river and coast is smallest. *A. japonica* starts the migration into the rivers in late autumn to early winter when the water temperature is 8–10°C and remains for 7 to 8 months. Environmental factors, beside water temperature, such as, tide, sunshine and wind influence their migration to rivers.

Once the eels start living in rivers, lakes, or ponds, they burrow between rocks or into holes or into the mud during the day and will go out into the waters to feed at night. The amount of feed increases from spring to autumn and eels grow up quickly during this season. The eels feed on various kind of insects, small fish, shrimp and shellfish as natural food.

The male eel becomes sexually matured in 3-4 years, whereas it takes 4-5 years for the female to reach sexual maturity. Towards the autumn the matured eels become bluish black on their backs, light golden on their sides and light pinkish on their abdomens, go down the river and enter the sea at the estuary. Immediately after entering the sea, the parent eels reach particular spawning grounds and die soon after spawning and fertilization.

The capture of eel is based upon the trapping method using a knowledge of the eels ecology, and different kinds of traditional gears have been used in various regions since ancient times.

Eels dislike light and prefer to live in dark places. The fishing methods making use of the eel's habit of burrowing in narrow places, such as, holes or between piled up rocks set bush traps, stone traps.

The fishing methods to attract eels with a paste bait, a mixture of powdered larvae of silkworm and rice bran, making use of eels highly developed sense of smell are used.

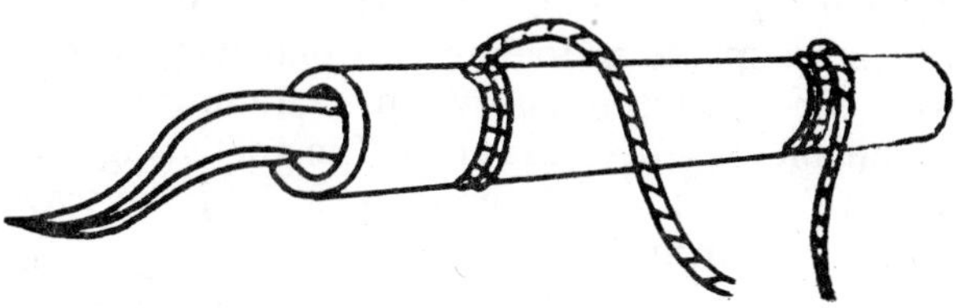

Fig. 99: Eel tube

Pot Fishing

The fishing grounds of pot fishery include coastal areas. Conger eel, cuttle fish, filefish, black porgy, and puffer fish are caught by pot fishing methods. Large scale pot fishing operations for crab and ivory shell have been conducted in the Bering Sea and the Sea of Okhotsk since 1960.

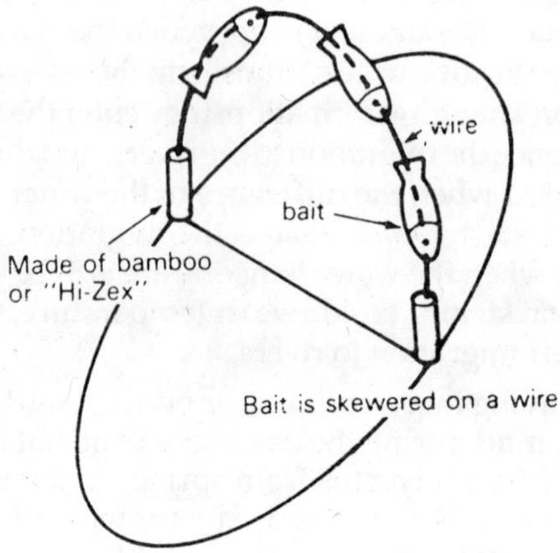

Fig. 100: Bait attached in crab pot

The modernization of gear and methods have enabled fishermen to take advantage of the following unique qualities of pot fishery as a fising method.

(i) Construction of the fishing gear is simple and the operation is easy to perform, and the introduction of motorized line haulers has enabled pot fishing to be conducted on a larger scale and with greater catching efficiency.
(ii) Pot fishing can be conducted in sea areas where trawl net and bottom gillnet fisheries are not possible.
(iii) New fishing grounds in the deep sea can be exploited.

There are three basic means of luring a marine animal into a pot; luring it with bait, stimulating its thigmotactic instinct, or providing it with a shelter-functioning pot. Among these, the most effective is the use of a bait.

In shallow sea areas with a depth of less than 50 metres, the sea is inhabited by a large number of marine animals, and food (bait) is plentiful. At depths more than 50 metres, however,the variety and number of marine animals decreases rapidly and the overall biota becomes sparse. So the presence of bait in a pot is thought to have an especially strong effect on the predatory demersal animals, such as, shrimp, crab and shellfish that inhabit these deep sea areas. The pot fishing method can be thought of primarily as a fishing method aimed at catching the scavenger-type predators inhabit such sea bottom areas.

Observing the actions of the marine animal that approaches the pot, it is evident that in addition to the animal's feeding instinct, there are also thigmotaxis and hiding instincts at work.

Especially some types of eels and octopuses show a definite preference for dark holes or crevices and the pot clearly functions as a shelter for those species. The catching effectiveness of the pot, in other words, its ability to hold an animal that has entered it in a trapped state, the size and location of the entrance (funnel) the effect of the shape of the entrance on the difficulty for a trapped animal to escape, the size of the mesh and the holding capacity of the interior compartment of the trap are all important factors, but difficult to make quantitative analysis of. Therefore, fishermen depend mainly on their own experience through trial and error in designing their own pots.

The following points are taken into consideration in designing a pot:

(i) Portability — It is important that the pots be such that a large number of them can be transported to the fishing ground at one time. Therefore, they should be designed either as collapsible pots or ones which can be stacked easily on top of one another on the deck of the fishing boat.

(ii) Lightness/Compactness — The pots should be light and compact enough to enable rapid sinking and hauling during the fishing operations.

(iii) Stability — It is essential that the pots have sufficient stability to assure that they land on the sea floor in upright position and that they are not easily knocked over on their sides. To achieve such stability, a stable cone-shaped design is often used, as is a metal frame on the bottom of the pot to serve as ballast.

(iv) Durability — The durability of pots has improved considerably with the use of synthetic fibre netting, but with square shaped pot the tendency for the netting at the edges to be torn or worn through is rather high. That is why, today, round pot shapes are preferred over square ones.

(v) Simple construction — Pot should be made of inexpensive, easy-to-find materials and designed in an easy-to-construct way. Also damaged pots must be easy to repair.

(vi) Use of large number of pots — In order to ensure an increased catching capacity it is necessary to set large number of pots over a large area, and for this purpose long line method is used. Also, with the addition of a line hauling device it is possible to fish in areas from 200 to 300 metres in depth for deep sea shrimp (prawn) and crab.

With regard to crab pots, two basic prototypes are recognized, one used for catching cold sea crabs and the other for catching warm sea "gazami crab". There is a basic difference in the position of the pot entrance depending on whether the crab to be caught is a "crawling" crab, such as, tanner crab or horse hair crab, or a swimming crab, such as, gazami crab. For a crawling crab the entrance is located on the top of the pot and for the swimming crab it is located on the side.

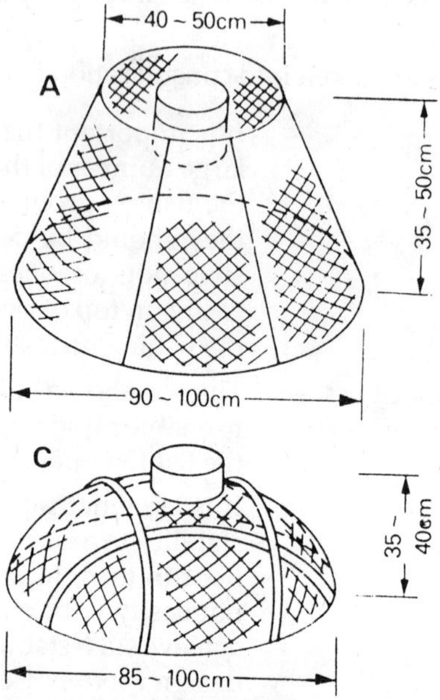

Fig. 101: Crab pot

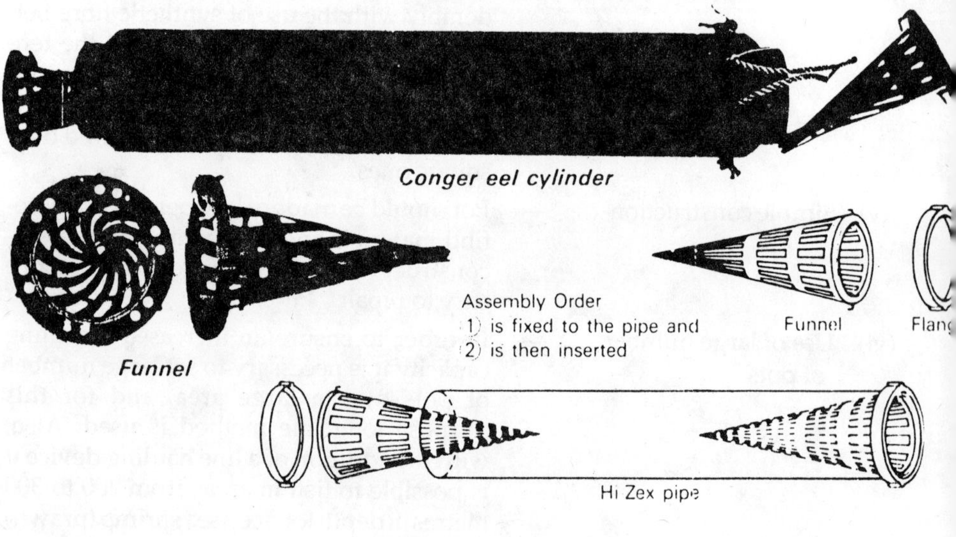

Fig. 102: Conger eel pipe

Three main types of edible shrimps are caught in 200-250 metres depth areas either in coastal waters or offshore waters by pot fishing. 5 tonne class boats are most commonly used for pot fishing inside the bay of 40 metres depth with flat sea bottom; while 9.9-tonne class is commonly used in open sea outside the bay.

The type of shrimp pot in use have the following characteristics:

(a) The round frame at the bottom of the pot is made of steel rod, while the upper frame is made of vinyl tubing.

(b) Cremona netting with 23 mm. mesh size.

(c) The entrance for the shrimp at the top of the pot is made by hanging a plastic cylinder; its colour is pink, which the shrimp are said to prefer.

(d) Hooks are fitted on the lower edge of the plastic cylinder for attaching the bait.

(e) Until 15 or 16 years ago the dimensions of the pots were 60-70 cm. in diameter at the bottom and about 50 cm. in height, but it was changed to shallower fan shape with a wider bottom ring diameter. These improvements were aimed at giving the pot greater stability on the sea bottom, while also making it easier for the shrimp to enter the pot.

(f) The netting on the bottom of the pot is fitted with an opening for removing the shrimp, which is closed by means of a draw-string.

(g) The netting on the top and slope section of the pot is stretched tightly over the frame. However, the netting on the bottom of the pot is left slack to allow the pots to be stacked easily on top of one another.

The pots are set in the fishing ground by means of a longline method. On one long line of 915 metres, 70 pots are attached and 7 such long lines are operated by each fishing family. The fishing operation is usually performed by a crew of three, but it can be operated by a crew of only two. The operation involves, (1) hauling the pots, (2) removing the shrimp, (3) baiting the pots, (4) closing the draw-string on the bottom net opening, (5) stacking the pots on deck, and (6) re-setting the pots. The process of hauling and resetting one long line of seventy pots require about one hour. The pots are left in the fishing ground for a period of three to four days. For bait, cut pieces of Atka mackerel, *Osmerus dentex* and saury pike are used. The fishing boats leave port before dawn and return by afternoon.

Eel Pot Fishing

Using a cylinder-shaped pot to catch conger eel is practised in the Pacific coastal waters of central Japan. The fry of conger eel (*Astroconger myriaster*), *Leptocephalus* appears in large numbers in coastal waters throughout Japan in early spring and grow to maturity in sand and mud sea bottom areas, where seaweeds grow in abundance in bay waters.

Mature conger eel live in waters with mud or sand and mud bottoms at a depth of greater than 10 metres and are found primarily in areas with a depth of 50–100 metres, although they can be caught at depths of up to 250 metres in some areas.

A "Hi-Zex" pipe cylinder pot was found to be superior, based on the following conditions, in addition to its catching capabilities:

Ease of water drain during the hauling operation.

Weight and ease of handling.

Volume and transporting (stacking) case.

Damage resistance and durabilities.

The "Hi-Zex" pipe cylinder-type conger eel pot is made up of the following parts:

1. The cylinder — In order to attract the conger eel, which dislikes light, with a dark hole, black Hi-Zex pipe is used for the cylinder. The pipe diameter is 15 cm., length 80 cm. and thickness 8 mm. A number of small holes, 4 mm. in diameter are drilled in both ends of the cylinder to serve as water drainage holes and also to let out the smell of the bait.
2. Funnel — A funnel-shaped entrance, which allows the conger eel to enter the cylinder, but prevents it from escape is fitted to each end of the pipe. At first the funnels were made of bamboo, later, however, an improved polyethylene one was developed.
3. Sinkers — In order to provide the cylinder with stability on the sea bottom, two old iron rods of 9 mm. diameter are attached to the cylinder by wires with a slight interval in between.
4. Branch line — A two metre length of 4.5 – 6 mm. diameter branch line attached to one end of the cylinder is used to secure the pot to the main-line.
5. Main line — An old tuna long line is used and the cylinders are attached to it at 25–45 metres intervals (depending on the water depth) and 30–60 cylinders are attached to one length of long line. A 3 to 5 tonne boat is used to operate three lines (a total of 100–180 cylinders) in one day.

In conger eel pot fishing, two or three frozen sardines or the entrails of other fish are put in the cylinder as a bait before it is lowered to the sea bottom. After one day the cylinder is hauled again and the catch is removed. The fishing grounds for conger eel are usually located at a depth of 50-100 metres.

In the cold season, they are also found in deeper waters at a depth of 200-250 metres in areas of mud or mud and sand bottom. The fishing operation involves raising the cylinders from the sea bottom by means of a line hauler, removing the lid holding the funnel in place, taking out the catch, placing in new bait, replacing the funnel and lid and stacking the cylinders on deck. When a full long line is hauled, instead of laying it again in the some fishing ground, the line is taken to a new ground for re-laying.

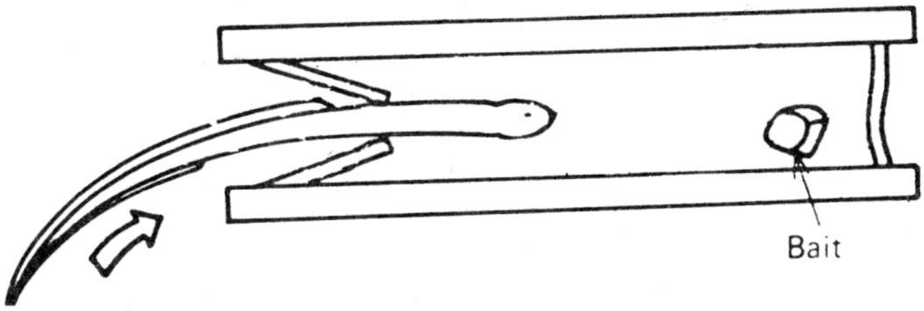

Fig. 103: Eel pot with bait

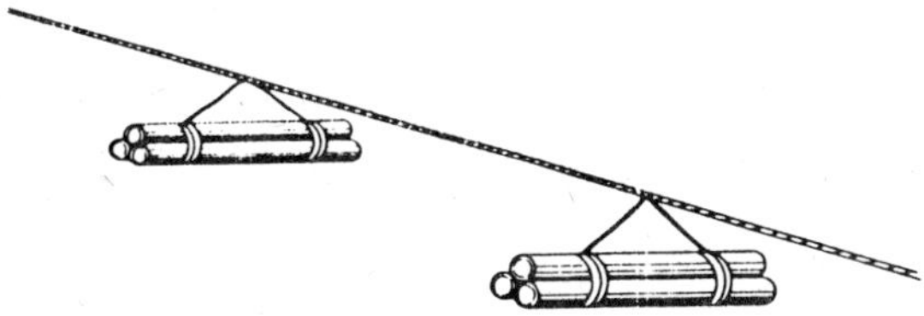

Fig. 104: Eel pot fishing

Pot Fishing for Assorted Species

Pot fishing method was successfully adopted for catching assorted tidal fishes. There are two advantages of this fish pot which make it effective as a commercial fishing method:

1. Unlike gill net this fish pot does not damage the fish and the fish is kept alive even if the pot is left in the water for several days, making this a flexible fishing method suitable for operation even during the busiest fishing season.
2. Because the fish caught by this method are landed alive, they bring a higher market price as fresh fish.

The pot is made of an anti-deteriorant treated polypropylene netting stretched over a stainless steel rod frame (3/8 diameter). The front of the pot is fitted with a funnel function entrance, while the back is equipped with a bag opening for removing the catch. On the sides and top of the pot the leaves of cryptomeria (*Raphiolepis umbellata*) or moss are stuck into the netting to darken the interior of the pot. It is believed that the contrast of light and shadow created by these leaves is effective in attracting fish to the pot. Also it is recognised that small shrimp will often attach themselves to the leaves, which is believed to increase the bait effect of the pot and no bait is put into the pots.

The pots are set in rocky reef areas with a depth of 30 to 60 metres, or in the narrow crevices between the reefs. The pots are left in the water for about 3 days. In a standard-sized pot as much as 11 kg. of fish, or as many as 40 fish have been caught. In the larger type of pot, large fish such as grouper are often caught, resulting the catches of up to 23 kg. at one time. The main species of fish caught by this method

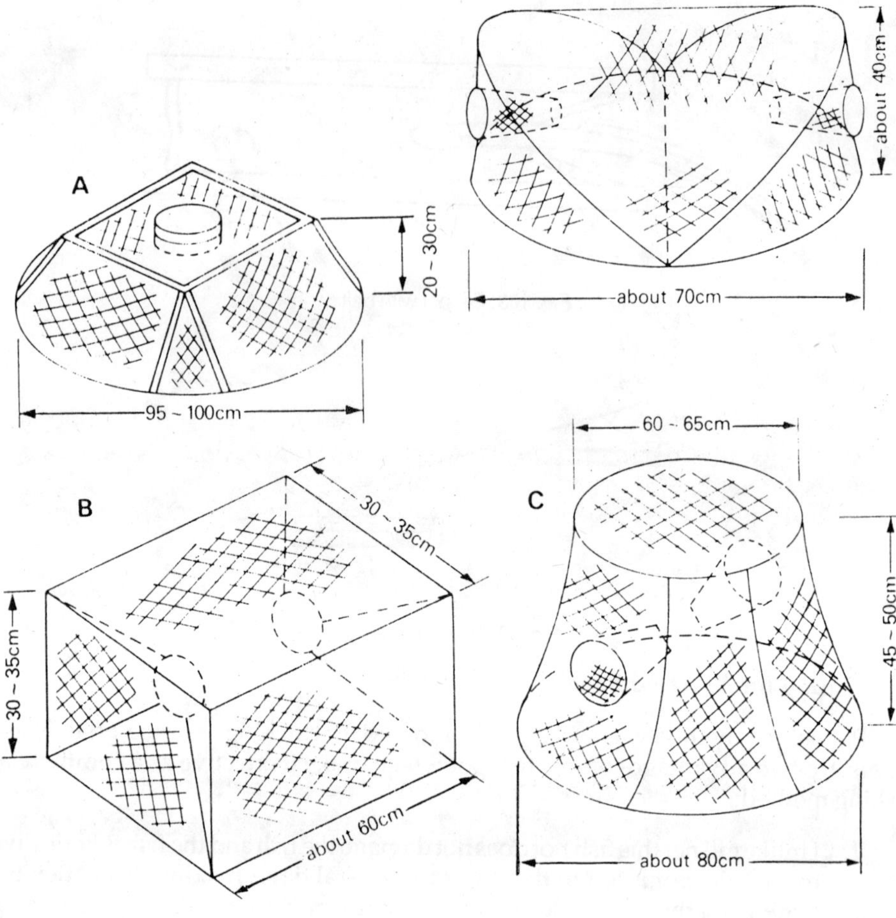

Fig. 105: Shrimp pots

include black porgy (*Pseudupeneus chrysopleuron*), Koshio bream (*Plectorhynchus cinctus)* and file fish. The number of pots that can be laid in one operation depends on the deck area and the stacking capacity of the boat. A 5-tonne class boats can carry 10 of the standard-sized pots or 5 of the large-sized pots.

Crab Pot Fishing

Gazami (*Portunus trituberculatus*) is a warm current crab which inhabits inland sea or bay waters, found usually at depths of 5 to 30 metres; and migrating between shallow and deep waters with the seasons. Gazami comes to shore waters of 5 to 10 metres to lay eggs, and at other times, is believed to retreat to a depth of greater than 10 metres. To catch Gazami crab of marketable size of about 15 cm., gazami pot fishing is conducted in the months of June to November, in sandy and mud bottom fishing grounds at a depth of 15 to 30 metres. August and September are the peak fishing season. Good fishing grounds are located near river mouths with inflowing fresh water or on the sea floor encircling a reef.

Compared to the crab gillnet fishing method, crab pot fishing has the two advantages of requiring less hand labour than freeing each enmeshed crab from a net, and it catches the undamaged ones, so that the catch can be sold as live crab.

In recent years, the collapsible type of standard pot is used in commercial crab fishery.

A brown netting stained with coal-tar is used for these pots, and the standard netting used is made of either nylon or polyethylene mono-filament. The size of the mesh is 12 cm.

The bait is skewered on a piece of wire, which is set inside the pot in a bow shape to attract the crab. For bait gizzard shad or small mackerel is used. The baiting of the pots is done before the fishing boats leave the port, and the pots are stacked on deck in collapsed state. On arriving the fishing ground, the pots are erected and thrown into the water one by one. The pots are hauled again several hours after setting. Between 60 to 100 pots are attached to one long line, and about 4 lines are set in one fishing operation. It is to be ensured that the pots rest in a stable position on the sea bottom and always fresh bait is to be used. The crabs are caught in the live state.

Crabs are scattered over a wide area. Pot fishery, which is a process of setting up pots in an area which seems to be the home range of crabs and catching them by luring with bait is easier and offers better selectivity than methods which involve towing nets around over a wide area.

By varying the mesh size and mouth size of the pot, it is possible to control the size of crab being caught. In order to conserve the reproductive capacity of the crab population, the body length of catchable crabs is restricted (90 mm. carapace width) and the capture of female crabs is also prohibited.

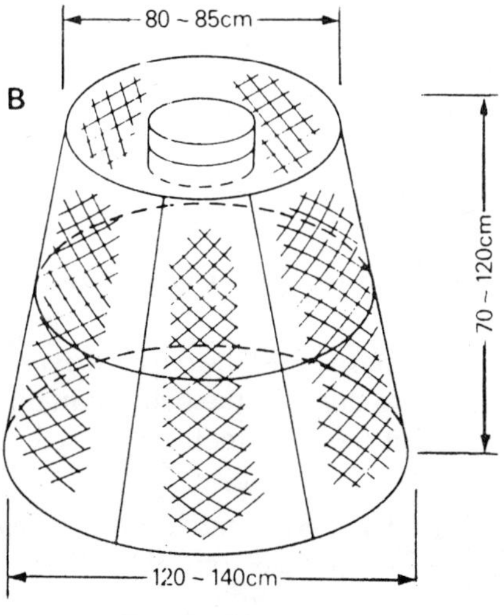

Fig. 106: Gazami crab pot

Shore Weir-weir

A place is selected where the river bottom is rather steep and the river is dammed. Two converging embankments are then built so there is a slight drop in the flow. In this way the flow of water is constricted. A screen of woven bamboo is then placed at the end to stop the passage of the fish. The fish land on top of this screen can be taken as captive. The shape and scale vary somewhat depending on the kind of fish and area.

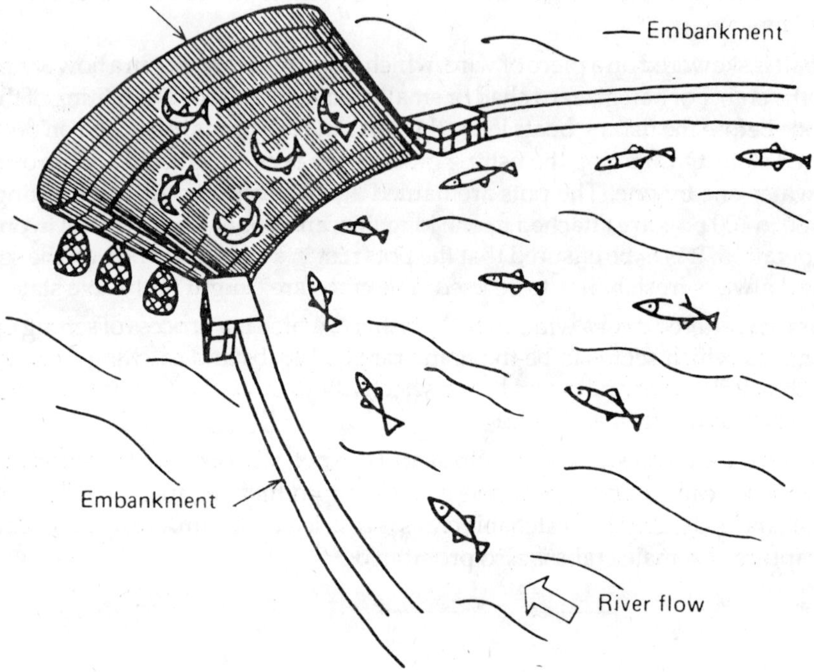

Fig. 107: Shore weir

Spear Fishing

In this method, large fish, such as, tuna, sword fish, shark etc. are speared from the front of the boat.

While the catch is still alive, the barb of the dart is driven in with 200–300 metre rope attached to allow the fish to dive. As the fish gradually weakens, the rope is slowly drawn in. For large creatures, this method is used on whaling ships with harpoon guns.

Octopus hooks and spears are used to catch octopuses hiding in the shade of rocks by hooking or piercing. When the octopus is hiding in a hole, it is lured out with a shore crab fixed to the tip of the hook.

Dive Fishing

Shallow coastal waters are important habitats for fishes, shellfishes, seaweeds

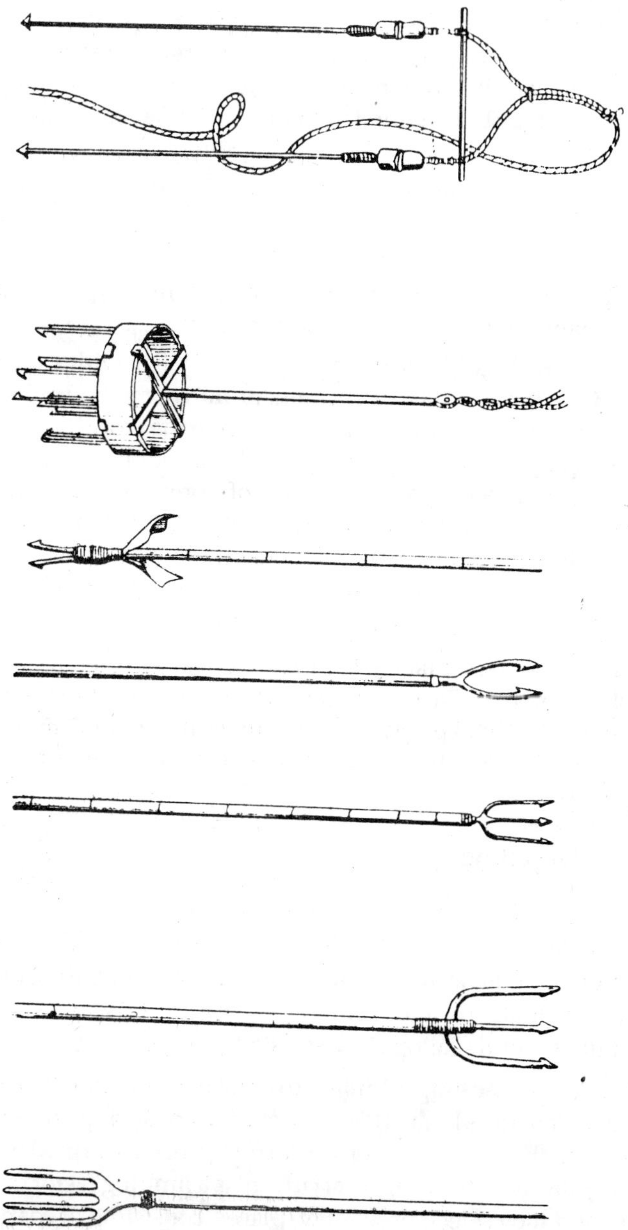

Fig. 108: Various Types of spear

and crustaceans. Shellfishes, seaweeds and crustaceans are sedentary or only slightly migratory. The biota of shallow coastal waters is rather complex, as the waters are varied, ranging from brackish waters affected by the inflow of fresh land water to high salinity waters open to the outer sea. Fishing grounds in shallow coastal waters may be classified broadly into "inner bay fishing grounds or inshore fishing grounds" and "open sea fishing grounds"..

The "inner bay fishing grounds" normally have a sea bottom consisting mainly of sand or mud; while "open sea fishing grounds" consist usually of rocky seaing bottoms.

The different environments result in different types of domnant marine resources inhabiting their respective fishing grounds.

Dive fishing to collect shellfishes or to catch fish with a spear without the aid of special diving apparatuses is an old and primitive method, which is now very rarely practised. It is practised only in a few places in the world, and is mainly confined to the circum-pan-Pacific belt area, e.g. the Tuamotu Islands, Arafura Sea, the Batan Islands of Bashi Strait, Japan and islands of Korea. They use simple fishing gear and gather seaweeds shellfishes and other sedentary fish and marine animals in the shallow waters of reef areas. In many areas, they are mainly female divers, but in some areas, male divers are engaged in this type of fishing.

Dive-fishing is possible only for 4 to 5 months during the summer.

Boso coast of Japan facing the Pacific Ocean is noted for a number of scattered dive-fishing villages. Divers collect seaweeds and shellfishes in the coastal waters diving April to October. There are female divers and male divers, the former normally work close to the beach and the latter diving in deeper reef areas employing boats.

Shellfish Harvesting

Scallop is a cold-sea bivalve, and its natural breeding places in the Japanese islands are distributed in (1) the Boso Peninsula and northward on the Pacific coast and (2) Voto Peninsula and northward on the coast of the sea of Japan. Especially in the coasts of Hokkaido and Aomori Prefecture (Matsu Bay), good fishing ground for catching natural scallops have existed since old times.

The habitat of scallop is limited to coastal waters shallower than 40 metres. The population density is high in the bottom areas consisting of coarse bottom materials. It is observed that scallops lie on the top of the ridges formed by tidal-currents. This seems to indicate that strong tidal current is a limiting factor or local distribution of scallops, and they select, preferably, places, consisting of coarse bottom materials.

For catching scallops, a small-sized power trawler equipped with shell dredge net is used. In the fishing ground, the net is thrown into the sea and towed at a slow speed of a few knots (2–3 knots). While towing forward, the net creeps along the bottom and sandy mud is scraped up by iron teeth fitted at the lower part of the mouth of net and scallops are caught in the net bag. Net hauling is conducted with a drum-type net hauler driven by the main engine.

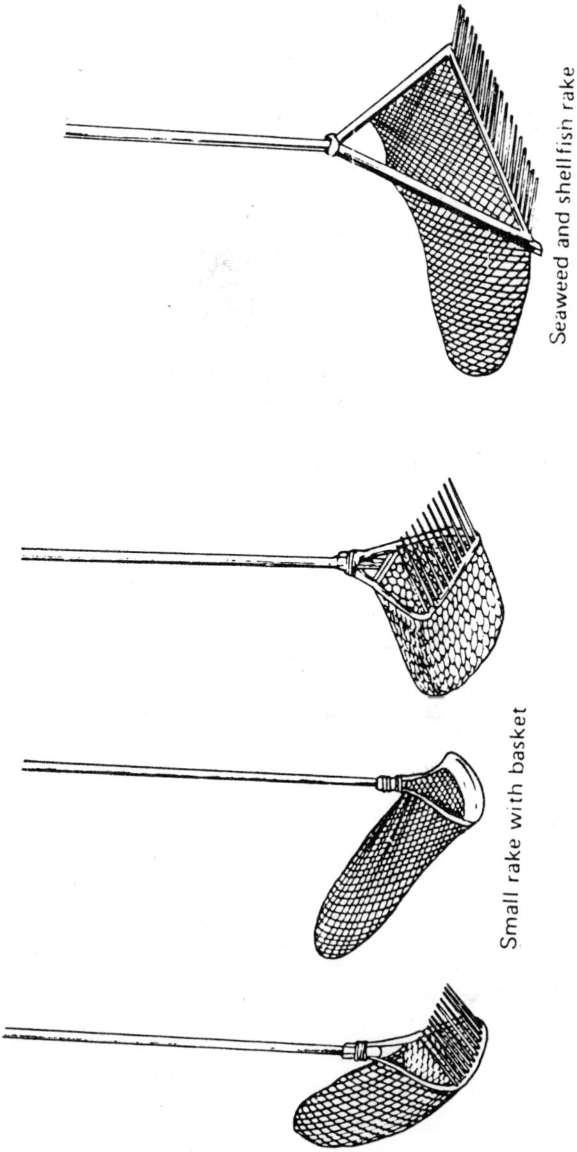

Fig. 109: Shell fish rakes

For conservation and propagation of resources, fishing methods (fishing season, operation area, daily catch limit, minimum shell length caught etc.) are to be restricted.

Harvesting of Seaweeds

Seaweeds are used as foods in two different ways; they are either eaten as they are or agar (poly saccharide) is extracted from the body of the seaweed and supplied to the food industry. Representatives of useful seaweeds used directly for food are of three kinds—green, brown and red algae. The species of sea weeds currently used for food in Japan are as follows:

Green algae	—	*Monostroma nitidum*
		Ulva pertusa
		Enteromorpha linza
Brown algae	—	*Heterochordaria abietina*
		Nemacystus decipiens
		Endarachne binghamiae
		Laminaria japonica
		Undaria pinnatifida
		Hizikia fusiforma
Red algae	—	*Porphyra tenera*
		Grateloupia filicina
		Carpopeltis angusta
		Eucheuma muricatum
		Meristotheca papulosa
		Gracilaria verrucosa

Forms used for food as well as processing form are divided roughly into the following three types:

1. Raw seaweeds — seaweed *salad* or spices for dish.
2. Dried products — Cooked and seasoned food and ingredients of miso-soup.
3. Processed products for food — Seasoned laver, tsukudani, condiments, confectionery etc.

As on the land, also in sea, depending on the region, depth, and conditions of the tidal currents, various aquatic plants adopted to each habitat grow abundantly. An important condition for the growth of seaweeds is photosynthesis and the existence of seaweed is controlled by the quantity of radiant energy from the sun. Almost all species of seaweed grow in water shallower than 50 metres, and especially shallower than 20 metres. Only a few species and a small number of seaweeds can grow on the seafloor at 100–200 metres depth. Among seaweeds, green algae grow in the shallow water, red algae in the deep water and brown algae in intermediate depth between the first two.

Other important conditions for growth, next to light are water temperature, salinity, waves, emergence during low tide, and substratum for attachment.

Harvesting of natural seaweeds are done by walking about beaches or intertidal rocky shores and by gathering on from a small boat at fishing grounds in shallow seas. In regular fisheries, the later method is employed.

Lavers of genus *Porphyra*, which are growing attached to the rocks exposed to the sea spray are plucked off by hand or with a small tool. Kelp and sargasso which are torn off by waves and drift to the beach are gathered up with rakes.

From shallow seas, gathering of seaweeds are carried out from a small boat operated by one person. A water glass is used to observe the bottom and seaweeds are pulled up by entwining the basal portion with a simple tool. Various shapes of gathering tools have been devised from old times in various regions.

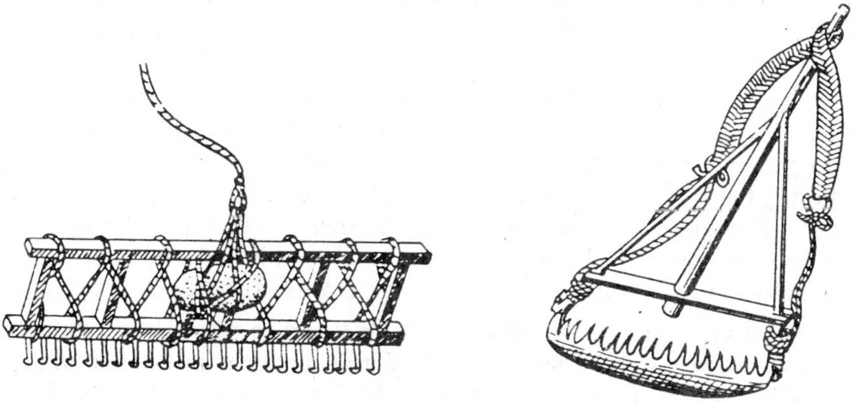

Fig. 110: Sea Weed harvesting rakes

Dried seaweeds contain 10–20 per cent water. The main components of these dried products are carbohydrates, such as, saccharides, and cellulose, proteins,

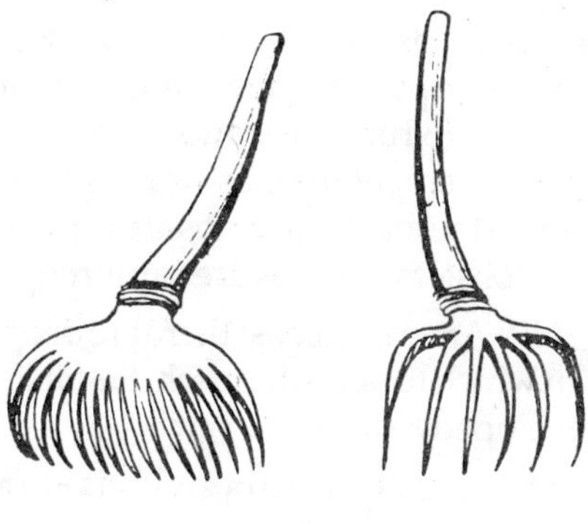

Fig. 111: Sea weed rakes

minerals, lipids and vitamins. The protein content of dried laver is 35–40 per cent of its total weight, which is nearly equal to that of soybean. Almost all kinds of seaweeds are rich in vitamin A, B_1, B_2, C and niacin. 7–34 per cent by volume of minerals are obtained from seaweeds. Minor elements, necessary for human body, such as, calcium, sodium, magnesium, potassium, phosphorus, sulphur, iodine and iron are also contained in sufficient amount in seaweeds.

Trap Boats and Potters

Small fish traps are used all over the world to catch lobster, crabs, shrimp, octopus, eels and all kind of reef fish. A small boat using 20 or 30 traps and landing 20 or 30 kilo of fish or shellfish per day, may make a profitable living. Some small boats set only 5 or 10 pots or traps, and many large modern vessels set a hundred traps or more at a time.

Pot fishing boats can work in any depth from 2 fathoms to 200 fathoms or more. Only a large well equipped vessel can operate traps at depths over 50 fathoms. A strong winch or line hauler is needed to bring the gear up from such depths. Lobster trap boats have fished successfully in over 200 fathoms depth in the canyons off the east coast of the U.S.A.

Most small trap boats have their wheel houses forward. They are usually open boats or half-decked vessels. The pot hauler is positioned on the starboard side near the steering position. A lead block is hung from a davit at the forward starboard rail. When hauling, the pot line comes in over the lead block and down to the line hauler or capstan. While one man tends the capstan, another lifts each pot aboard as it comes up, and takes out the fish caught. Hauling is accomplished with the vessel's head to wind. A steadying sail is sometimes fitted aft to help keep the head to wind.

Setting of pots is done over the stern or side. As the traps should be set close to rocks or crevices, the use of a paper-recording ecosounder is invaluable in the operation. When working very close inshore, landmarks are used as the best position indicators. Trappers working offshore require radio direction finder, loran or other navigating equipments to determine the location with precision.

Some trap boats are high speed vessels (20-30 knots), which are expensive to operate.

Most lobster boats work day trips only and do not use ice. Trap boats working for fish and those making longer trips use both ice and live tanks in which to keep the catch. The tanks need to be aerated or cooled and circulated regularly. Care must be taken to keep the tank water temperature as near as possible to the seawater temperature at the seabed where the lobster or crabs are captured. A deck wash tank is normally used for cleaning lobsters or crabs are captured.

Off both the NE and the NW American coast there are deep sea trap boats which use refrigerated live wells in which to keep the catch. Suitable small offshore boats with live wells might be developed to keep the catch.

Shrimp can also be caught in pots or traps, on grounds too rough for otter trawls or where trawling is prohibited.

11

DECK MACHINERIES OF FISHING BOATS

To make up a well designed and efficient fishing boat, innumerable fittings and pieces of equipment are required. This includes warp leads, rollers, gallows blocks, derricks, gilsons, tackles, booms, cleats and hawse pipes. For efficient performance, every activity on a fishing boat should be studied properly, paticularly the practical problems of life and work on the vessels before design and construction.

Masts and Derricks

The oldest and most basic installations on boats, mast or mast-like structures are vital if all the necessary lifting operations are to be performed. Wooden masts are still perfectly acceptable, or if available, steel pipe is even better. With steel, one can fabricate bipod or tripod masts which can leave the centre foredeck clear, and can take greater loads than a simple wooden mast. After-masts or mizzen masts are more and more being made of steel and built in to the superstructure along with "A" frame side derricks, rader masts and antenna poles.

Blocks and Pulleys

Blocks and pulleys are attached to strong points on the masts and derricks. Good blocks should be made of best handwoods, or nylon or steel, and should have proper bearings or bushes that are self-lubricating or can be lubricated. For boats with small winches or limited power the 3 fall and 4 fall blocks can enable the crew to lift heavy loads on board.

Fairleads

Rope and warp leads should be positioned with great care. They should be strong, hard-wearing and have enclosed ball-bearings or lubricated bearings on which to revolve. Rail leads are normally portable. The rail is fitted with inset pipe brackets and supports at intervals and the rail lead is fitted into these as needed. The rail lead normally has one horizontal and two vertical rollers. A chain or nylon rope is attached to the lead and is connected to a ring or eye fitted under the rail. This prevents the roller from being pulled out in heavy weather. As far as possible, warps and warp leads should be clear of the fish and net working areas. They may require protective covers to prevent injury to crewmen standing nearby.

Power Take-offs and Winch Drives

From the power take-off point, belt drive is preferable to chain drive and by menas of jocky pulley can be adjusted for different loads. Vessels with hydraulic installations require only a Vee belt drive to the pump. This must be geared to produce the required r.p.m. speeds. A clutch should be fitted at the power take off point. In many situations hydraulic drive may prove to be cheaper than mechanical drive in the long run, provided that expertise and spare parts are available locally.

There is a limit to the number of items that can be driven from the forward end of a marine propulsion engine. Alternatively a powerful auxiliary engine is used to drive winch, pumps, generators and other deck gears.

Capstan and Line Haulers

These are the smallest and simplest winches available. Even the smallest boat may now be fitted with a capstan or line hauler capstan are normally slow-turning and line haulers usually fast turning. Capstans generally have the range of 270–1000 kg. at speeds varying from 60 to 300 feet per minute. Line haulers fall in the same power range but have higher maximum hauling speeds. Heavy duty capstans are used in purse seining or as gipsy drums on trawl winches. Capstan heads should be made of hard wearing material. Iron, steel, bronze and other alloys are used.

The rope must lead straight on or at an angle away from the outside or open end of the capstan. The crewman must lay on just enough turns as are needed to take the load and no more.

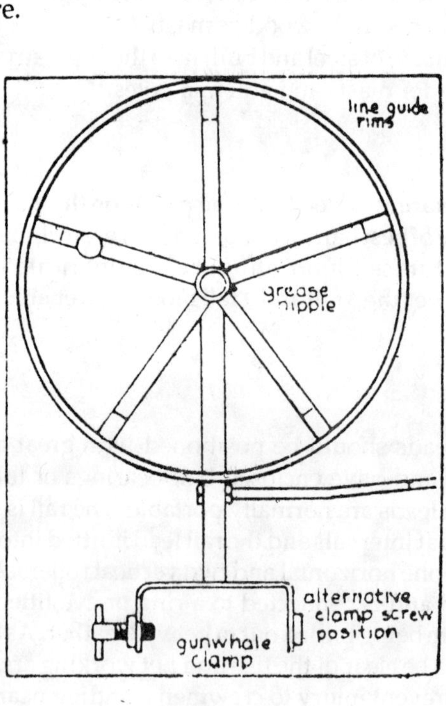

Fig. 112: Gurdy

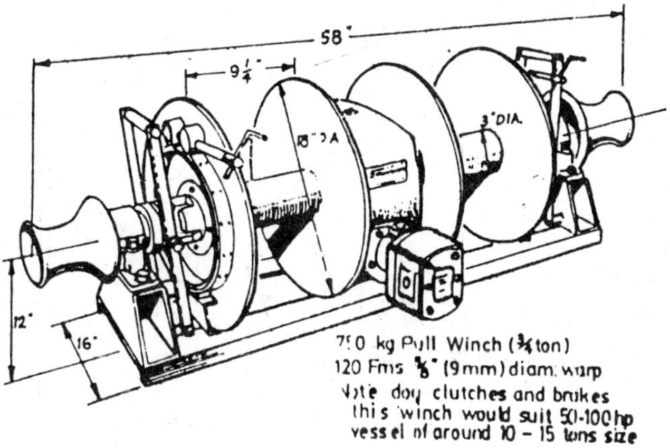

Fig. 113: Winch

Fishing Reels

A larger version of the reel one finds on angler's rods, used chiefly for snapper and other large line fishing. The reel can be used for single line reef fishing or single line trolling. Reels can be hand operated, or powered, either electrically or hydraulically.

Gurdies

Small line haulers used to haul in trolling lines, hand lines, and jigging lines. The line is wound on to the gurdy instead of just being pulled through. Gurdies are designed for specific fisheries such as, salmon trolling, mackerel lining or cod jigging. Gurdies reduce the amount of labour requred to tend several lines at the same time. Most powered gurdies are hydraulically operated. Some can be driven by simple hydraulic systems runoff an out board engine. The lines run from each gurdy through blocks or rollers extending from booms or over the rail.

Jigging Machines

The best known jigging machines are used to catch squid. They can be manually operated or driven by hydraulic or electric motors. Hand operated ones consist of a simple wooden drum with handle, and a lead leading out over the rail. The fisherman creates the jigging motion by erratic movements of the drum. Oval

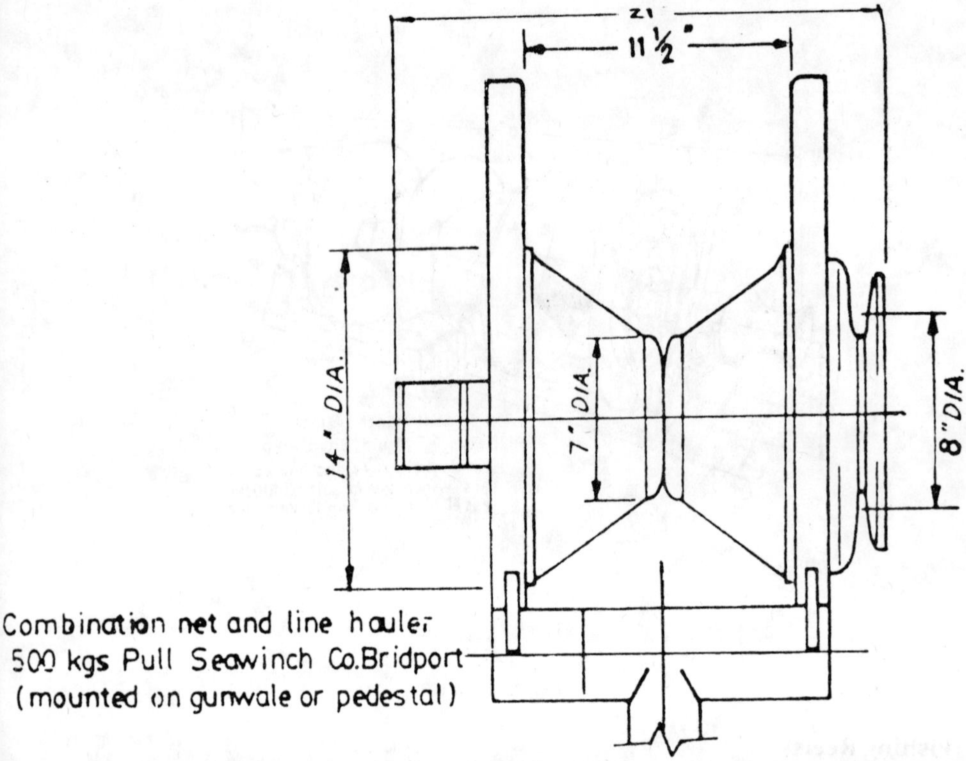

Fig. 114: Net and line hauler

shaped drums are used to create the jigging motion on units which are mechanised and run off a regular speed drive. A large boat may have seven or eight pairs of jigging drums on each side of the vessel.

Winches

All trawl and seine winches are composed basically of one, two or three drums on to which the wire ropes are wound. Their hauling speeds are usually slow as the strain on the warps is normally quite high. Winches with operating speeds of 10 to 90 r.p.m. would have hauling speeds of 10 to 200 feet per minute.

Trawl winches can be single drum (split type), two drums on shaft (the conventional type) or two or more drums on parallel shafts (shrimp or purse seine types).

The basic components of any trawl winch are the main drive unit with the worm gear; the main shafts and bearings, the clutches; the brakes; and the guide-on gear. A winch may or may not have a capstan at one or both ends.

Both dog clutches and friction clutches are used on trawl winches. Dog clutches are strong and effective; while friction clutches are gentler that they do not transmit the loads so suddenly on to the winch. Boats with heavy duty deep sea trawls prefer dog clutches on their winches, while those using light trawls or midwater gear generally prefer friction clutches.

Winch brakes are used only when shooting the gear to prevent the winch from over-riding the warp. When the warps are all set they are screwed up tight to hold them at that point. The brake linings have to be renewed from time to time as they wear rapidly with constant use.

Guide on or spooling gear is used to make the warp wind evenly on the drum. More sophisticated winches have automatic spooling gear fitted. Proper spooling reduces wear on the warps.

Drum capacity and warp size (diameter) and length should be accurately specified when ordering a winch. The specification of winch might be as follows:

Winch type	—	For 23 metre 350 h.p. wooden stern trawler.
Lay out	—	Two drums, single unit, with twin capstans.
Warp pull	—	2.5 tonnes at mid-drum, 5 to 50 metres/min.
Drum capacity	—	400 fathoms of 12 mm. wire plus 50 fathoms 18 mm. cable.
Clutches	—	Friction clutches
Accessories	—	Manual guide-on gear.
Drive unit	—	2000–3000 psi Hydraulic Pump, tank, motor, piping and controls to be included. Pump drive from 1000 r.p.m. power take off in engine room.

Hydraulic Drive

Simple hydraulic units involve a fixed displacement pump, an oil reservoir, filters and valve controls. The hydraulic motor is attached to the winch or power block itself. Multiunit installations usually require, variable displacemet pumps, which are expensive, but extremely useful in those situations. Low pressure systems are used on large vessels mainly. Medium and high pressure systems 1500–3000 psig (105–210 kg./cm^2) are used mostly by fishing boats of 30–120 tonnes, 100–600 h.p.

Net Haulers

Net haulers consist of a revolving sheave or drum or of two sheaves held together by a strong spring. They can be designed to suit any type of net but chiefly gill nets, trawl nets and purse seines. Most net haulers are hydraulically powered. They may be fitted to a bracket on the rail superstructure, fitted on a pedestal bolted to the deck, or hung from a boom or hydraulic crane.

Gill net haulers take several forms. Plain cylindrical drums are used to pull entangle nets and trammel nets. Sheave blocks like small power blocks are more common for ordinary gill nets. Normally they are mounted horizontally at or just above head height for the crewmen standing on deck.

Net drums are also used for gill nets, trawls and seines. They are large and powerful, and require a fair amount of deck space. Noramlly a rail roller should be fitted in front of the drum. Drums may be mounted at the stern or on the side deck or foredeck. Most drums are fitted near the stern. The main advantage of a net drum is that it reduces the labour required in setting and hauling the gear. It also reduces the possibility of the gear snagging or being fouled when setting.

Power blocks are used chiefly for purse seine hauling, but also for hauling light trawls, mid-water trawls, ring nets and large drift nets. The most common block is the single sheave type or "puretic" power block, but double sheave, triple sheave or triple roller types are also available. Single sheave blocks come in sizes of around 15 inch, 19 inch, 24 inch, 28 inch and 32 or even 36 inch diameter. The purse seines used by the 60 feet 120 h.p. junking boats in South East Asia measure around 200-400 metres long and 30-60 metres deep. They could all be handled comfortably with a 19 inch power block. For a similar sized net a north Atlantic boat would need a 24 inch block. The pulling power of these blocks is more than adequate for most fishing boats. The net should be allowed to slip when the tension is excessive, or else the hauling speed should be reduced.

Wheel House Equipments for Fishing Boats

Fishing vessels require position fixing equipment for safety and navigational purposes and for fishing operations. It is possible for small open boats to fish successfully without electronic equipments, because of the seaman's professional and experienced eyes for constant visual scanning for land, ships, or lights. Fishermen can also detect schools of fish by observation of the sea surface and the sea birds or by inference from the state of wind. Old fishermen use lead lines to determine not only the depth of water, but also kind of sea bottom by placing grease in the depression at the bottom of lead, where some mud, sand, gravel or shells would stick. Lead lines wire are also used to determine the density of schools of pelagic fish by experienced fishermen. Fish striking the wire as they swam past would set up vibrations which by their frequency could indicate the size or density of the fish school.

Position Fixing

To obtain a "fix" on its position, any ship or vessel at sea requires at least two position lines. The two lines should be at as near right angles at each other as possible for the sake of accuracy. There are many ways of obtaining these position lines. Traditional fishermen determine them from visual observation of hill top, houses, trees, points of land and other immovable objects. They are computed mentally with no chartwork necessarily being involved. Compass bearings are the next most common type of position line. Radio signal bearings are another form of compass bearing, as are celestial position lines from sextant observation of stars or heavenly bodies. Hyperlbolic position lines are obtained from radio navigational systems, such as, Loran, Decca, or Omega. Satellite systems are also now available.

The Magnetic Compass

One of the most important instruments a fishing vessel should have. Magnetic compass is simple, reliable and efficient and require practically no maintenance. Proper marine compasses are fixed on gimbals to compensate for the rolling of the vessel. The compass card is balanced inside in an alcohol or similar liquid. Smaller compasses for little boats are simply bolted to the bulkhead. For large vessels a binnacle compass with compensating iron balls are used. A sight vane for taking compass bearing is very useful. It may be removable. A dummy compass card with sight vane is also helpful. Overhead compasses are very useful for the helmsmen.

Radio Direction Finders (R.D.F.)

Radio direction finders are used to determine the direction from which a particular radio signal is emanating. They are normally tuned to radio stations which can be identified by their special calling signals. But they can also be used to locate any vessel transmitting radio messages. They are of use mainly to boats fishing in areas where other systems do not exist and on voyages of several hundreds of miles.

Radar

While radar is an excellent device for detecting other boats at sea or when making a landfall, it is not a very useful instrument for actual fishing. They are available with maximum ranges varying from 10 to 60 miles or more. The rotating antenna is fitted to the wheelhouse roof or to a special radar mast.

Loran, Decca and Omega

These are the best known of the radio navigation grid systems which provide constant coverage from a series of master and slave stations on hand. None of the systems give world-wide coverage and one must select the best available for each region. The range and accuracy of each is also different. Decca with an approved range of about 200 miles is the most accurate, followed by Loran which has a longer range. That is followed in turn by Omega which is for ocean-going vessels only. Fishing fleets in the north-east Atlantic use chiefly Decca and those on the north west use Loran.

Fish Detection

The best known and most widely echo-sounder is used for position-fixing and sea bed identification as well as fish finding.

The recording echo-sounder traces an echo graph of the seabed under the vessel on a roll of paper of 10, 15 or 20 cm. wide. Wet paper contains an iodine or chemical that turns brown when an electrical impulse is passed through the paper. Dry paper has a carbon interior and the surface paper is burnt off by the electric impulse from the stylus pen.

The frequency at which a sounder operates will determine to a large extent the quality or sensitivity of the recording. Low frequency (10–40 kHz) give good ground discrimination (sea bed picture) and can detect school of fish, but not fish without airbladders (mackerel, squid, tuna). High frequency echo-sounder (80-200 kHz) are very good at detecting fish without air bladders, but they do not give good ground discrimination. Mid-frequency sounders (40-80 kHz) have a performance somewhat between the other two.

Bottom trawlers normally use low frequency sounders to give them a good picture of the sea bed. Tuna and mackerel purse seiners, squid jig vessels and pole and line boats would use high frequency machine. Handline and gill net boats might prefer a sounder with mid-frequency transmission.

"White line" effect is produced on some sounders to help detect fish schools lying close to the bottom. For detecting fish lying close to the seabed, deep sea trawlers now use the scale expansion and bottom lock sounders which give a picture of 5 fathoms or 10 metres immediately above the seabed.

Eco-sounders are usually available in scales of up to 500 fathoms or more. Small inshore boats may have echo-sounder with a maximum range of 50 fathoms (100 metres). Many vessels are fitted with two echo-sounders, one for shallow water and one for deep water operation, or one for low frequency sounding and one for high frequency operation.

Net Transducers

Net headline transducers or "netszonde" are fixed to trawl nets to determine the mouth opening and/or the depth at which the net is fishing if it is a midwater trawl. They are simply plugged into a paper recording unit for display purposes. The cable between the transducer and the ship is wound on to a special net sounder cable, designed for the purpose.

Sonar

Sonar is a horizontal scanning echo-sounder. Its transducer is fixed to a pipe which can be lowered or raised from within a larger cylinder located near the keel towards the fore end of the vessel. The transducer itself, when lowered, can be turned to either side of the vessel, sometimes to any direction in 360° Sonar transducer may also be trained up or down, some even from +10 degree to –90 degree. Other sonars may have a fixed beam angle or allow for the selection of one of three angles. Most sonars are paper-recording types, but several models offer the option of ppi screen display. This is easier to read in relation to ship's head or course direction but it does not provide a permanent record.

Most sonars are used in purse seine fisheries to locate schools of pelagic fish. But they can be used in bottom trawling to locate patches of hard bottom before the vessel or net has reached them. Some gill netters use sonar when working close to rocks or wrecks.

Sonar sets are equipped with audio "speakers" which pick up the echoes and convert them to sound at a level we can hear. An experienced skipper with a well tuned ear can detect fish echoes by listending to the sound.

Fishing Net Making and Mending

The net is constructed fundamentally by the meshes. Therefore, the meaning of net making is to compose many meshes in certain number of depth and length. There are two ways of net making, namely, machine making and hand making. A mesh composed of 4 knots and 4 bars.

A mesh is the smallest unit of a sheet of webbing from which commercial fishing nets are constructed. It is a rhombic opening enclosed by four bars of twine of equal length firmly knotted at the four corners. Each corner is called "Knot".

There are many different types of knots used in net construction. The following indicates different types used in webbing:

- Webbing
  - Knotted
    - Sheet Bend
    - Double Sheet Bend
    - Reef knotted
  - Knitted or Knotless
    - Knotless
      - Straight type
      - Zig-zag type
      - Tortoise type
    - Minnow
      - Ordinary type
      - Twined type
      - Woven type
    - Bussel

Sheet bend features firmness, minimal slippage or loosening and feedom of mesh stretching movement in any direction is generally employed. It is effective for gill nets and entangle nets.

Reef knotting is usually employed in Asian countries. This has the defect of being relatively easy to loosen, but it makes for easy knotting. Hydro dynamic resistance of this netting is less than the sheet bend and therefore used in purse seine.

Knotless type, produced by twisting two strands and then inter-twined at intervals to form mesh without knots have greater tensile strength than knotted counterparts. As this contains less twine, it is less bulky and consequently the hydrodynamic resistance is smaller. But it has a defect of being difficult to repair when torn. Recently, this netting has been used extensively for set nets, purse seines and trawl net in Japan.

Usually the edge with clean meshes of webbing is called "length", while that with cut meshes is called depth.

Webbing size is measured by depth and the length. The depth of webbing is indicated by either (a) number of meshes or (b) number of rows. The length of webbing is shown by either (a) number of meshes, or (b) metres, feet, yards or fathoms. In the latter case, the length is to be measured by stretching the net lengthwise.

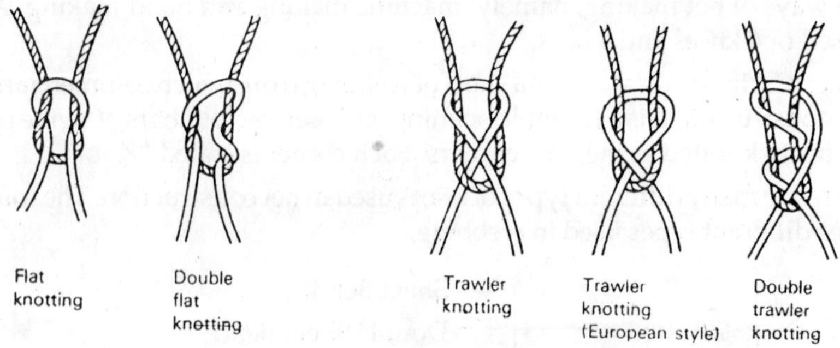

Fig. 115: Different type of knots

Method of indicating mesh size

Stretched measure	All countries	Length of two bars.
Square measure	Some Europan countries	Length of one bar.
Pasada	Spain, Portugal	No. of meshes per 20 cm.
Row	Holland, England	No. of rows per yard.
Fushi or Setsu	Japan	No. of knots in 6 inches.
Omfar	Norway, Sweden	No. of half mesh made in one "alen" (one alen is equivalent to 0.628 m. in Norway and 0.594 in Sweden.

The assembling of pieces of webbing to make up a net and connecting them to the ropes is called "hanging".

Proper hanging is essential for the right degree of fullness in all parts of the net, proper relief of tension by the selvedge and, on the whole, the intend performance of the gear in water and its fishing success.

Making the Knot

The implements used for hand braiding of netting are needle and spool. Pass the needle up through the mesh until a short tail of twine remains below the knot. Using the fingers, twist the tail of the twine around the two bars of the mesh above the twine leading to the needle. Throw the loop of twine to the left to form a sheet bend and pull tight. Thus a half mesh knot is made. It is better to mend nets quickly and efficiently. The rows of knots must be lined up in the right direction and the netting properly supported.

When making half mesh knots from left to right across the net the needle is brought up through the mesh. The size of the loop (half mesh) should be formed by adjusting with the finger. The thumb and index finger clamp the twine at the centre point of the half mesh on which the knot is being tied. A sheet bend is formed by making a loop to the left and the needle is passed through a half mesh over the loop. Then the knot will be formed by pulling the twine tightly with the moving of needle to the right.

When working from right to left the needle is passed down through the mesh. The palm of the hand is towards the net making operator. The twine is clamped in the half mesh by the index finger and the thumb and a sheet bend is formed in the same way.

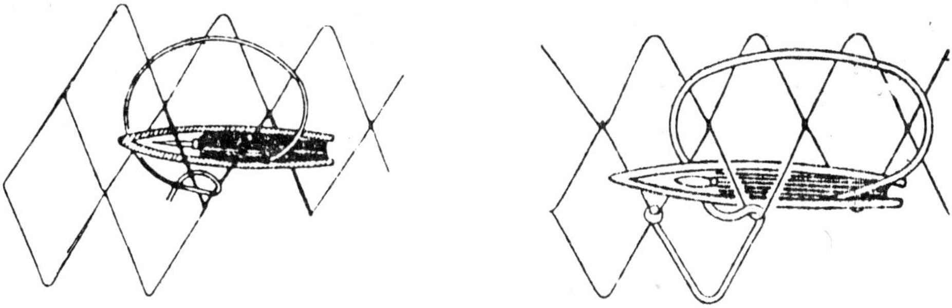

Fig. 116: Making knots

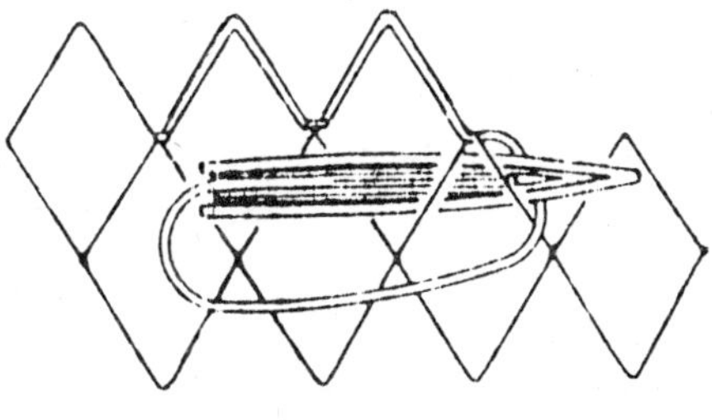

Fig. 117: Making Knots

Right Side Knots

Lay the twine to the right of the knot of the side mesh with the index finger behind the knot. Measure the length of the bar of the mesh being formed by lining up the two knots above the one being made. Clamp the twine with the index finger and thumb just below the knot in the side mesh. Throw the loop in the twine to the left from an overhand knot around the knot in the side mesh and pull tight by moving the needle to the left.

Left Side Knots

Lay the twine to the left of the knot in the side mesh with the index finger behind

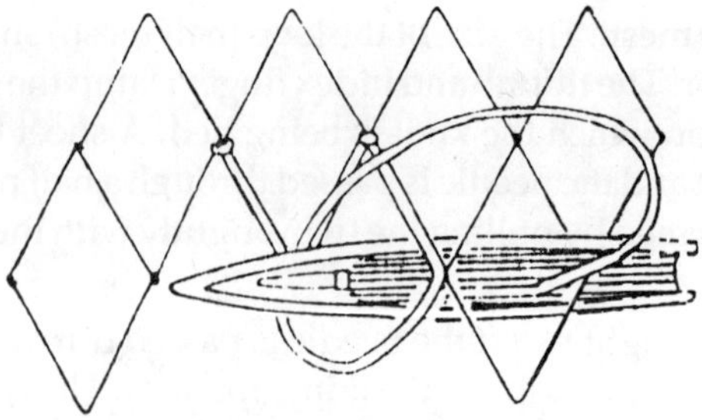

Fig. 118: Right side knots

the knot. Measure the length of the bar in the same manner as for the right side knot and clamp the twine in the same way. Throw the loop of twine to the right from an overhand knot around the knot in the side mesh and pull tight by moving the needle to the right.

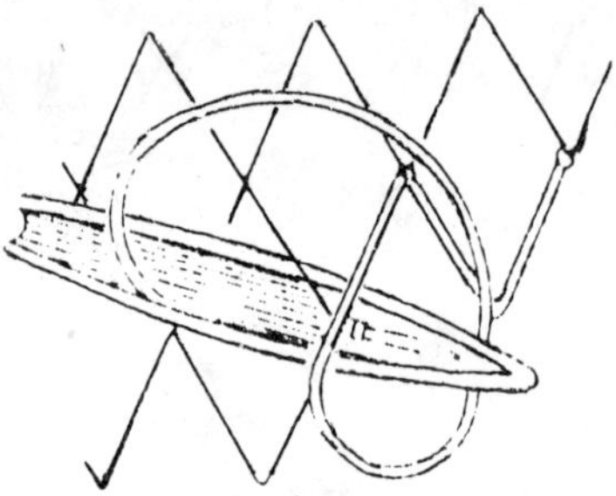

Fig. 119: Left side knots

Basic Fishing Net Composition

(a) Net braiding can be done by machine or by hand. Any size of mesh can be braided by the net machine. Also several types of knot nets can be braided with various thickness of twines, as flat knot net, trawler knot net, double trawler knot net, knotless knot etc.

(b) Kinds of knot— Flat knot (or reef knot or square knot) and trawler knot are most popular type of knot in practice.

Joining:–When two different compositions of net pieces are joined together, the case is to be considered as follows:

(i) Two net pieces are of same length but different mesh size.

(ii) Two net pieces are different in length but of same mesh size.

(iii) Two net pieces are different both in length and mesh size.

Lacing

The selvedge runs along each side of the net strengthening the gear. The selvedge is to be made of the twine of the same size as that used for the webbing. To lace the selvedge of two net pieces together, for instance the belly of a trawl net, three or five meshes of each net piece are laced with a needle of double twine making two round turns and a half hitch at a distance of one mesh for the full length of the selvedge.

Combination and Joining the Webbing

Several pieces or units of netting are combined into a set of net. In the combination, there are two main ways of joining the webbing; the sides of net pieces are joined to increase the width or the edges of net pieces are joined to increase the length.

Net Mending

The damage to metting gear includes the breaking of netting and the cutting of twines used for joining the webbings, ropes etc. In case of small damage, that part only need be replaced or repaired with the netting twines of the same kind and size.

When working from left to right across the net the needle is passed up through the mesh. The length of the bar of the mesh being formed is measured by lining up the knot being formed with the adjacent knot on the left. Clamp the twine with the index finger and thumb. A sheet bend is then formed by throwing the loop of twine to the left and passing the needle under the half mesh being picked up and over the loop. The knot is then pulled tight by moving the needle to the right.

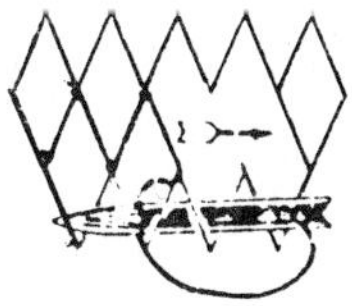

Fig. 120: Net mending

When working from right to left the needle is passed down through the mesh being picked up and the length of the bar of the mesh being formed is measured by lining up the knot being formed with the adjacent knot on the right. The twine is clamped between the forefinger and thumb and a sheet bend formed, throwing the loop in the twine to the left and tightening by moving the needle in the left.

12

NEW WAYS OF CAPTURING FISH

Electrofishing

The use of electro-fishing equipment to sample fish populations in fresh water lakes and rivers has been an accepted and useful practice for many years. The equipment employed is nearly always logically designed and great variation in techniques and performance are reported (Myers, 1951; Rollefson, 1958; Burnet, 1959 and Patten and Gillaspie, 1966). To a large extent this wide variation in equipment and technique is a result of the great variability of water conditions, fish species and sampling needs in freshwater lakes and large rivers coupled with the fact that almost any electrofishing system will allow capture of some fish under conditions suitable to the particular system.

Electric seine with alternating current A. C. system have been used by Biswas (1961) for clearance of fresh water ponds from predatory fishes and aquatic animal to use the ponds for fish larvae and juvenile rearing. Capture of fishes from large fresh water ponds have been made successfully by using the same electrofishing system (Biswas, 1970).

Basically, the function of an electro-fishing system is to produce a sufficient electrical stimulus in fish near the electrodes to permit easy capture by netting. In an A. C. system, the electrical stimulus simply immobilizes fish (Electronarcosis or

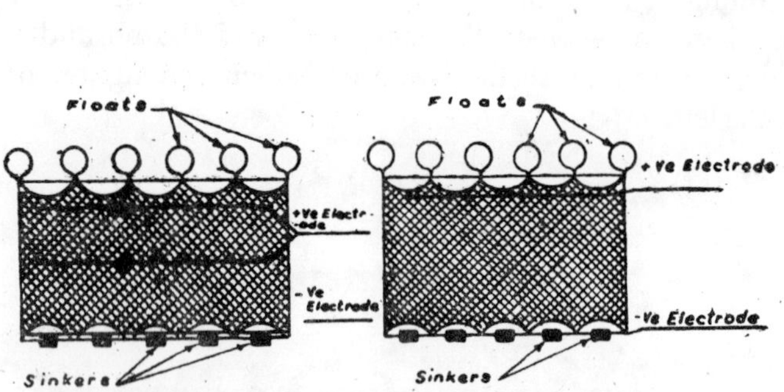

Fig. 121: Electric seine devised by Dr. K.P. Biswas, 1961

electrotetanus) requiring the net handlers to collect fish from considerable distances or depths.

Often fish are immobilized at depths beyond reach of the dipper and are not susceptible to capture at all. In direct current (D.C.) or pulsed direct current system, fish near the anode exhibit forced swimming (electrotaxis) toward the anode and hence more easily netted.

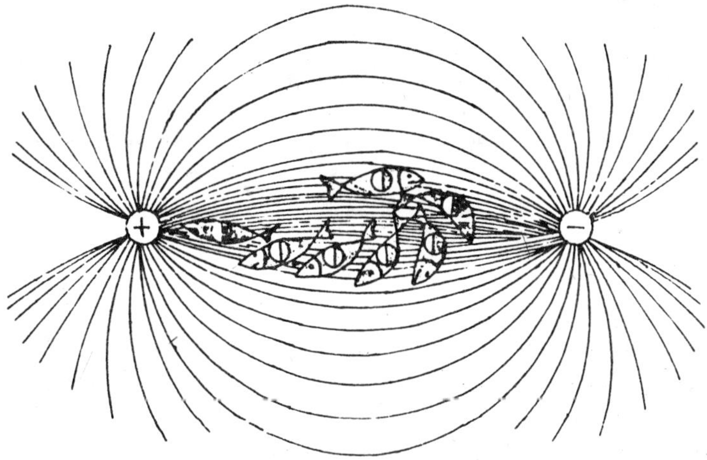

Fig. 122: Forced swimming of fish towards the anode (galvanotaxis)

It is quite well established that fish are much more susceptible to A.C. than to D.C. and hence the radius of action of an A.C. electrofising system operating at a particular voltage is much greater than a D C system at the same voltage. Pulsed D.C. is more effective than D.C. and to some extent combines the desirable forced swimming response of D.C. with the larger radius of action associated with A.C.

Based on this concept, for any electrofishing system there exists some minimum value of current per electrode which will effect fish out to some specified distance from the electrode. This minimum current will depend upon electrode shape, nature of current (A.C., D.C., pulsed D.C.), desired range of action, fish species, water temperature, water conductivity and many other less obvious factors. Of these factors, only electrode shape and type of current are under the control of the designer. All other factors are external variables which must be compensated for by variations in the operations of electro-fishing system.

The distance at which fish are affected can be too small due to incorrect electode arrangements and insufficient current (power).

Low water conductivity makes it very difficult to attain sufficient currents to produce useful electrofishing responses. Extremely high water conductivity calls for currents too large to be supplied by portable equipment without special electrical control methods. For very high conductivity D.C. is ineffective (Vincent, 1971) and pulsed D.C. or A.C. is more effective.

Voltage ranges for A.C. electrofishing

Conductivity (micromhos/cm.)	*Voltage (V)*	*Power (KW)*	*Remarks*
10-20	460*	2.0**	* Denotes limit imposed
20-40	460*	3.5**	by equipment rating
40-60	460-390	4.5*	** Denotes limit imposed
60-100	390-320	4.5*	by maximum electrode
100-160	320-230	4.5*	size
160-350	230-160	4.5*	
350-700	160-100	4.5*	

Turbid water and excessive vegetation restrict visibility and reduce the value of immobilizing capabilities of A.C. electrofishing. Pulse D.C. and D.C. can solve this problem.

Fish which are immobilized at depths exceeding 0.9–1.2 m. (3-4 ft.) are very difficult to capture by ordinary netting procedures. Electric seine (Biswas, 1970), pulsed D.C. and D.C. can help to overcome this problem.

High conductivity bottom material tend to "short circuit" the current out of the water into the bottom material reducing the electrofishing effectiveness and sometimes overloading the power source. Avoiding direct contact between electrodes and the bottom may help to remedy this situation.

Theoretically, success of electrofishing should increase with the increase of temperature, as the conductivity of water and water temperature has more or less a straight line relationship (Sigler, 1969). But workers in the field have noted differently.

Mortalities caused by A.C. electrofishing are higher than those caused by D.C. or pulsed D.C. (Taylor et al., 1957). Mortality is usually associated with regions of excessively high current density close to the electrodes.

Individual variation is notable among fish even though they are of the same species and have similar lengths (Haskell et al., 1954). The larger the individul of a species, the more sensitive it is to a given electric shock (McMillan, 1928; Mclain and Nielsen, 1953; Taylor et al., 1957). Fish absorb power as a function of body surface area and particularly length (Holzer, 1931). Also the greater resistance of smaller salmonids (Nakatani, 1954) and possibly small fish of other species as well, further reduces their response to shocks. All of these factors contribute to capturing a greater proportion of large fish that are actually present in the population.

Some species are quite difficult to capture by electrofishing. The most notable are northern pike and muskellunge. It appears likely that this is caused by the strong swimming ability and natural tendency to escape danger by rapid swimming coupled with a high sensitivity to electric fields.

Indian catfishes followed by bottom dwellers are found to be more sensitive to A.C. shocks than those species live in surface and mid-layers (Biswas, 1961 and 1970).

Large differences in response have been observed between various fish species when pulsed D.C. is used. Some selectivity in electrofishing may be possible because of the differences. There may be optimal wave forms and pulse rates which will produce selective and efficient electrofishing for various species.

Numerous problems associated with equipment limitations and lack of operating guidelines exist in electrofishing operations. There is a need for developing guidelines for electrofishing system capable of adjusting for variations in external conditions. Although the design and operation of effective electrofishing system is still to some extent an art because of lack of basic data on fish responses to various levels and types of electrical stimulation, there are a number of basic concepts which are useful in guiding the development of new electrofishing equipment.

The components of a fixed electrode electrofishing system can be classified into four sub-systems according to function :

1. Boat and mechanical — to effectively and conveniently carry the complete system.
2. Electrical — to generate, control and deliver the electrical energy to the electrodes
3. Safety — to provide proper safeguards for operating personnel, and
4. Electrodes — to properly couple the electrical energy to the water.

Boat and Mechanical Components

Selection of the boat, motor and other mechanical components be made as to convenience, availability and ease of maintenance. Metal boats offer the advantage of simple reliable grounding of all electrical equipments through the physical attachment of the equipment to the metal boat. Fibreglass or wooden boats would, however, require careful grounding practices to assure electrical safety.

Electrical System

In addition to the main generator, the electrical system of an electrofishing boat should include an auxiliary generator and battery, metering, disconnect switches and overload protection, voltage controls and power conversion equipment, and a safe and reliable system of interconnecting the various components. A means of rapidly disconnecting the power in case of emergency must also be provided as a part of the safety system.

Main Generator

A three-phase A.C. generator provides the greatest flexibility. In A.C. electrofishing all three phases can be effectively utilized by proper electrode design. If D.C. or pulsed D.C. is desirable, a transformer rectifier system provides a voltage controlled D.C. source without need for large filters as would be required with a single phase generator.

Auxiliary Generator and Battery

Use of an automotive-type, 12 volt D.C. system for auxiliary power has been proved very effective. The entire system consisting of generator, voltage regulator, meters and battery can be integrated with the main generator to form a compact self-contained unit.

Meters

For A.C. electrofishing, one voltmeter and a set of three ammeters is sufficient. When transformers are used, it is convenient to place the ammeters in the primary circuit. In additon to a voltmeter in the primary circuit to monitor generator performance, a second voltmeter in the secondary is useful to indicate the actual output voltage.

For D.C. operation, a voltmeter and an ammeter are essential. In case of pulsed D.C., two voltmeters are useful, one to read peak voltage and one for average voltage.

Overload Protections

As in any electrical system, a disconnect switch and overload protection must be provided, since in an electrofishing system the electrical load vary considerably because of changes in water conductivity. For this circuit breakers are much preferred over fuses for protection of the main power circuits.

Voltage Control

The use of transformers to control the output voltage and thus provide a proper match between electrode resistance and the generator characteristic is extremely valuable for operating over a wide range of water conductivity.

Safety Systems

For safe operations, equipments must be selected and designed to avoid potentially hazardous situations. Power connectors having metallic parts, which could become energized through insulation failures and net handles which could form an electrical connection between the water and an operator must be avoided.

A low-voltage system should be used for all electrical functions except the actual electrofishing function.

Proper grounding practices to assure that all metallic parts in the boat are joined electrically are absolutely essential.

A rapid disconnect system is essential. Crews must be aware of the hazards associated with electrofishing equipment and be properly trained in safety and rescue procedures.

Electrode Systems

The requirements of an effective electrofishing electrode system include:

1. Establishment of an effective electric current distribution in the water to be used.
2. Avoidance of local regions of unnecessarily large current densities, which waste power and are potentially harmful to fish.
3. Adjustability to meet changes in water conductivity.
4. Ability to negotiate weeds and obstructions.
5. Ease of assembly and disassembly.
6. Avoidance of unnecessary physical disturbance to water to permit easy visual observation of fish.

The critical parameters in any electrode system are:

1. Each electrode is surrounded by a region of steadily decreasing current density which normally includes a "danger zone" close to the electrode, an "effective zone" at moderate distances and a "perception zone" at lager distances.
2. Large electrodes (cross section dimension) have three desirable properties:
   (a) lower resistance, hence more current at a given voltage;
   (b) a larger "effective zone" because of the increased current density at moderate distances;
   (c) a smaller "danger zone" because of increased surface area and resultant lower maximum current density.
3. Electrode separation is a relatively unimportant parameter whenever the separation is large compared to the cross-section dimension of the electrodes.

The relative importance of electrode dimensions depends upon the specific electrode. Spherical and thin cylindrical electrodes offer boundary cases which are useful in estimating actual electrode performance. An electrode in which all dimensions are of the same order of magnitude will behave similar to a spherical electrode (rings, square plates), whereas electrodes having dimensions which are orders of magnitude apart will behave similar to a cylindrical electrode (long thin cylinders, long thin plates).

For sphreical electrodes:

(a) The resistance is dependent on electrode radius and varies in inverse proportion to this dimension.

(b) For separations, large compared to the radius, each electrode operates independently of all others.

(c) Increased electrode radius causes a lage increase in the "effective zone" and a large decrease in the "danger zone".

For cylindrical electrodes:

(a) The resistance depends upon the ratio of separation (D) to radius (R) and upon the length. The variation is very slight in terms of changes in the ratio of D/R and large with respect to changes in length (in inverse proportion).

(b) Increased electrode radius causes a moderate increase in "effective zone" and a moderate decrease in "danger zone".

(c) Decreased separation causes a moderate increase in both effective and danger zones.

(d) Increased length simply extends the field further into the water.

Spherical electrodes have generally superior electrical properties but have many mechanical disadvantages. Thus electrode arrays which achieve the benefits of spherical electrodes while utilizing the cylindrical elements are often most effective.

Operating Guidelines

General—

1. A conductivity meter to measure the water conductivity at the time of electrofishing operations are to commence on a given body of water is essential. Knowledge of the water conductivity will allow the operator to adjust the voltage source and change electrode size to conform with the conductivity to ensure maximum current densities during operation.
2. Metering devices are essential so that the operator knows that sufficient currents are being provided.
3. Operation at highest allowable current produces maximum effectiveness whenever possible 80 per cent of this value to allow for normal conductivity variations be adjusted.
4. Large electrodes and low voltage operation is superior to operation at higher voltages. Higher voltages be used only when low water conductivity makes it necessary.

5. The safety switching system can be used as a control to turn the power on and off. This can be effective in approaching areas of possible high fish concentration without power on and then energizing at appropriate time. The net handlers can exercise this control.
6. Knowledge of the waterbody topography and the habits and habit of the fish species sought along with some planning before the initial field work can improve the success of electrofishing.
7. It is generally more feasible to fish at night, especially for those species that move towards the shallows at night.
8. All electrical systems (power, safety and auxiliary lights) are in good working conditions to be ensured, especially the safety system, before operations begin.

A.C. Operation—

1. Fish are not attracted by A.C., but this type of current has the largest range of action, especially in clear, shallow water with a sand or gravel substrate.
2. Occasionally, better performacne is obtained at reduced output if fish are being stunned too far from the electrodes.
3. It is necessray at times to operate at high boat speeds especially when seeking northern pike and muskellunge.

D.C. Operation—

1. Fish will be attracted to the anodes without stunning. This can be advantageous when water clarity is poor, vegetative growth is dense, or algae prevents reasonable visibility.
2. The range of D.C. is minimal. Low-pulse rate D.C. is usually better than D.C.
3. It is desirable to operate at very low boat speeds to allow the fish to swim up to the anode and remain there until netted. When excessive boat speeds are utilized fish will be attracted to the anode but won't hold for any extended period as they quickly become exhausted, sinking or drifting back under the boat.

Pulsed D.C. Operation—

1. Low pulse rates (5–15 pps) are similar to D.C. except they have greater effective range.
2. Higher pulse rates (40-80-120 pps) have greater range but fish will be stunned as they approach the anodes.
3. Pulsed D.C. offers the same advantages as D.C. when operating in turbid water or dense vegetation.

4. As with D.C., slow boat speeds are mandatory to fish effectively.
5. Pulse duty cycle appears to have only secondary effects. Settings at 25 and 50 per cent appear to give similar results and hence 25 per cent is preferable since it conserves power. The 10 per cent setting appears to be less effective.
6. Some species selectivity is possible with pulse D.C. Although only qualitative observations have been made, the following may be useful.

   Trout, carp, bull heads and large mouth and small mouth bass respond well to higher pulse rates and will approach quite close to the anodes before beign stunned. Walleye, yellow perch, blue gills and white and yellow bass are more easily stunned and lower pulse rates are required to bring them close to the anodes. There appears to be an optimal pulse rate for each species.

Electrofishing and Safety

Electrofishing uses voltages and currents which can be lethal for the operator. Portable, electrical equipment used in moist, outdoor environment is more prone to fail and cause shock hazards.

20 to 500 hZ currents are more dangerous than direct current or higher frequency currents. The current which passes through a body that does the damage. The voltage in a circuit is only important insofar as it can produce current in the body. Large current passing through or around a person can cause serious injury because of the heating and burning of tissue. This type of injury is more common at higher voltages used by power companies.

Death at lower voltage, such as 120 to 240 V found around homes, can usually be attributed to one of three following causes:

(a) Ventricular fibrillation.

(b) Respiratory arrest.

(c) Asphyxia.

Ventricular fibrillation is an uncoordinated asynchronous contraction of the ventricular muscle fibres of the heart in contrast to their normal coordinated and rhythmic contraction. The heart seems to quiver rather than to beat. This condition is caused by an electric shock where the path of current is through the chest. Once a person goes into ventricular fibrillation, the only way to stop the fibrillation is to use a defibrillator which applies a pulse shock to the chest to restore the heart rhythm. Closed chest heart massage and artificial respiration may help until the victim can be defibrillated.

Shocks with a current path through the respiratory centre can cause respiratory arrest. The respiratory centre is at the base of the skull slightly above a horizontal line from the back of the throat. Thus, shocks from the head to a limb could lead to respiratory arrest. Artificial respiration can help in this case.

Asphyxia is caused by contraction of the chest muscles. When current is above a certain level, a person cannot let go of an electrically hot wire. Currents somewhat above this level may not be sufficient to cause ventricular fibrillation but may be sufficient to cause the contraction of the chest muscles and asphyxia since the victim cannot let go of the wire.

Operating Instructions for Electrofishing Safety

Warning— Electrofishing equipment is hazardous. In addition to the usual safety precautions are as follows:

1. Be sure both boom grounding clamps are attached to suitable grounding points in boat.
2. Learn the location of each off switch. Power to the electrodes can be interrupted by pressing any of the off switches or by unloading any of the mat switches.
3. Do not place hands or conducting objects in the water near the boat. There are hazardous voltage gradients in the water near the boat as well as near the electrodes.
4. Do not bypass or use dummy loads on mat switches.
5. Stop the generator before changing fuses, removing or inserting plugs, or making any changes in the electrical system or electrodes. The generator should not have to be operated to have lights available.
6. Although the 12-volt system does not pose a serious shock hazard, it is capable of large currents and associated arcing hazards. Do not treat these circuits and the battery carelessly.
7. Do not operate the A.C., pulsed D.C. electrofishing boat close to other boats or near a shore line where people or animals are located.

Electrical Ratings and Switching

Caution— Failure to observe the three items below can cause damage to the equipment:

1. The main power control relay is the gernerator control box and is controlled by the safety switches. Always disconnect the generator from the pulser by pressing one of the off switches before changing settings (voltage level or mode) on the pulser.
2. The main generator is rated 230 V, 11.7 A. Do not operate for extended periods with any of the three A.C. ammeters reading more than 11.7 A.
3. The 12-volt generator is rated 12V, 55A. This system is identical to that on an automobile and should be monitored as in an automobile by observing the 60-0-60 A ammeter on the generator control box.

Circuit Protection

1. The main generator is protected by a 15-ampere circuit breaker in the generator contorl box. This breaker functions as a manual on-off switch as well as the circuit protective devcice. It is reset by simply returning it to the on position after correcting the over load condition.
2. The 12-volt circuit is protected by a 60-ampere fuse and the control circuit by a 1.0-ampere fuse, both mounted in the generator control box.
3. A special fuse inside the pulser protects the pulser circuits aganst direct short circuits. This fuse cannot be changed in the field.

Operation

D.C. – (a) Set mode control to D.C.

(b) D.C. ammeters should all read nearly the same and preferably be in the range of 10 to 11A.

A.C. – (a) Set mode control to A.C.

(b) Be sure anode switch is on.

(c) A.C. ammeters will normally not all read the same. The largest reading must always be less than 12 A.

(d) In some cases it will be possible to operate at a higher voltage than in the D.C. mode. If the largest current is 8A or less, press an off switch and then raise the voltage. If the largest current is less than 12A, operation is acceptable.

Pulsed D.C. – (a) Set duty cycle switch to 50 per cent. Turn mode switch to desired pulse rate. The D.C. ammeter should now read approximately 50 per cent of the reading obtained with the mode switch on D.C. If it does not, the pulser is misfiring on this setting. Try other pulse rates or reduce D.C. current by adjusting electrode.

(b) With duty cycle switch on 25 per cent or 10 per cent the D.C. ammeter should drop to 25 per cent or 10 per cent of the reading on D.C. for any of the available pulse rate settings. Failure to read this levels indicates misfiring of the pulser.

(c) Misfiring of the pulser will not cause harm to the equipment but may not give good electrofishing results. Reducing the current by adjusting the electrodes will usually cure the misfiring.

(d) If the pulser should stop pulsing and go into the D.C. mode, while set for pulsing, try interrrupting the power momentarily by pressing an off buttton and then a start button. This type of malfunction can be caused by a stray pulse or switching transient and is not harmul to the pulser circuit.

Safety Procedures

Following safety precautions should always be observed:

1. Always wear Coast Guard approved life jackets.
2. Always wear hip boots or waders.
3. Store all gear and keep boat in neat condition.
4. Avoid excess fatigue or drowsiness and be constantly alert.
5. Authorize one person to be in charge and his authority should not be questioned during an operation.
6. Instruct all personnel in the fundamentals of electricity.
7. Thoroughly familiarize all personnel with all phases of the equipment and its operation.
8. Make sure that all equipment is in good condition and properly used.
9. Make sure that there is a first aid kit and fire extinguisher on the boat.

13

FISHING IN THE FUTURE

In order to get advantage of greater catch per unit time of actual fishing and to find and capture fish with more certainty than present-day gears, fishery scientists are in constant search of designing some such gears and equipments that would operate independently of the weather conditions at the surface and thus could fish more days each year.

Some techniques, which D. L. Alverson of U.S.A., Bureau of Commercial Fisheries Exploratory Fishing and Gear Research Base at Seattle and Dr. J. Wilimovsky, Director of the Institute of Fisheries at the University of British Columbia, Vancouver, considered might contribute to fish catching in the future are given below:

1. Determination of watermass boundaries and surface temperature by infra-red procedures.
2. Single-wavelength light beams might facilitate visual fish detection down to depths of 100-200 m. or even deeper.
3. There is a scope for refinements in sonar equipment for detecting the direction and speed of fishes by use of doppler sonar, and in high-resolution short-ranged equipment for identification of targets by studying three-dimensional patterns of fish school shapes characteristic of the species.
4. It might be possible to develop recording spectrophotometers for tracking organic odours in the sea and thus identify fish schools. This may lead to use of artifical odours to force or guide fish along a given path. Introducing water soluble chemicals into an air-bubble curtain might enhance its effectiveness as a fish barrier.
5. Reproducing sounds of prey or predator might be used to attract or herd fish. Electrical fields may be used in conjunction with light, sound or chemical stimuli to aggregate and lead fish. Retrievable floats with built-in detection system could automatically signal to catcher vessels the presence of fish.
6. A network of unmanned buoys will detect fish and transmit data through satellite telemetering to a shore-based "hydro-central" for

computer analysis and transmission of data summaries by facsimile technique to fishing centres.

7. Manned underwater stations for research and harvesting are entirely conceivable. The environment may be improved by fertilizing lagoons and by installing nuclear reactors at the sea bottom to create thermal upwelling. Plastics and light weight metals will be increasingly used in vessels and gear. Irradiation and other innovations in preserving seafood may affect the pattern of fishing operations.

Some of the above ideas are still just engineering dreams, but some have been built and tested, and all seem scientifically feasible.

A net of unmanned buoys has been established for several years in the sea and the patterns of occurrence and distribution of natural resources have been determined and plotted. The buoys are interrogated at regular intervals through satellite telemetering and from their surface transmitters by pulse-coded sonic means to instrument heads at various depths in the sea. Transmission redundancy is reduced to a minimum as only points of parameter change is telemetered.

As the data come into "hydro-central", computers reduce the mass of informational bits to contoured plots of biological oceanography and meteorological parameters. By facsimile techniques, these data summaries are transmitted to the research laboratories and fishing centres of the world. When a biological parameter anomaly occurs, the nearest buoy would automatically be instructed to assess the nature of the instance with high-resolution sonar and autospectrophotometric methods. These data would be transmitted back to hydro-central for computer and human interpretation. The movements of the identified resources would be plotted.

In some instances it might be necessary to verify the nature of the resources or an anomaly by an on-the-spot check using aircraft equipped with Laserscope or hydrofoil research craft equipped with high-speed self-propelled submersible television vehicles. Depending on the species, the main fishing fleet could be deployed into the path of the fish, or conversely, suitable deterrants could be placed in the sea to guide the fish to the catcher.

Aircraft could disperse the necessary chemical pellets to olfactorily guide the fish, or remote controlled underwater vehicles would produce the necessary electrical-sonic or bubble barrier to perform the same function. Depending on the depth of harvest, catches would be performed by catcher boats assigned to permanently anchored factory ships or by automated underwater vehicles operated from ship or shore stations.

Surveillance of the main plotting board in hydro-central would allow detection of weathers conditions and precursors of EL Nino type shifts in advance. Similarly, areas of high or low basic nutrient production could be watched with broad environmental limits, spawning populations deflected accordingly.

Such highly automated equipment should not only make it possible to catch more fish at lower cost and with less human effort, it may also open the way to the exploitation of new resources. In the past, when fishermen have used improved equipment or cast their lines in unfamiliar depths, they have often discovered unsuspected concentrations of fish. Some years ago, the ocean perch of the Pacific was considered a fish so rare it had no commercial importance. But when fishermen started using deep water trawls, they brought up so many perch that it is now among the most common food fish on the west coast of U.S.A. Experimental trawling at even greater depths in the Pacific has recently revealed the presence of huge schools of hake, a codlike fish that may some day prove to be a valuable commodity. In designing new equipment, fishery scientists have taken advantage of the ways that fish respond to what they see, hear, smell and feel. The dye curtain acts like a fence, because many species of fish are reluctant to swim from clear into turbid water. Supplementing the dye is a noise barrier; a net of sonar transmitters that emit a repellend sound, e.g. the noises that fish themselves make when alarmed by a predator. Effectively confined by the dye and the noise, the fish can be scooped up by a trawl towed by two submarine engines. A television camera enables the fishermen on the surface to watch the proceedings, and they guide the trawl by means of acoustic signals.

In front of the dye curtain there may be two permanently anchored data-collecting buoys. A third buoy, near the bubble curtain may float freely, meant to be retrieved by a ship. Instruments mounted on the buoys and hung on their cables gather information about water conditions, e.g. temperature, salinity and currents, and the figures are automatically telemetered to shore, where they can be processed by a computer and developed into synoptic diagrams, like weather maps, that indicate where fishing is likely to be good.

Very large nets often catch fish that evade smaller ones. An American research group has been testing a huge trawl, 317 feet in length. These scientists have found that some elusive species get so thoroughly lost in it that the net can be closed before the fish realise that they are in danger.

Fish can be led into the giant seine by an attractive-smelling substance strewn by the helicopter. Fish are astonishingly sensitive to odours. The helicopter might also be equipped with a laser-powered fish finder, as radar is not very useful since its beams hardly penetrate water. But light from a laser can probe down to several hundred feet.

The self-propelled trawl is designed to trawl at high speed. In front of the trawl, electrodes be fitted as protuberances which can stun fish so that they cannot escape.

Fish can also be confined by a curtain of bubbles that come from a compressed-air hose laid along the bottom. Bubble curtains, which work on much the same principle as dye and noise barriers, have already been used successfully by Marine sardine fishermen in shallow water. In front of the bubble curtain, a suction pump can pull fish up to the boat. Such pumps are now used in the Caspian Sea, and also by some men haden fishermen in the Atlantic. Self-propelled "beaters" can drive the fish to the pump by means of a pulsating electric field that causes fish to line up between the anode and the cathode. Fish of different sizes respond to pulses of different frequencies. Thus by tunning their beaters, fishermen can select the sizes they want to catch.

While reviewing and critically examining the ideas stated in the foregoing paragraphs, S.J. Holt and G.L. Kesteven are of opinion that some of the ideas, represented above are already beyond speculation. In the Caspian Sea, fish are attracted to fish pumps by lights; submarines are already used for collecting materials on the sea-floor; automatic instrumented oceanographic buoys have been developed.

The illustrated methods differ from today's in the extent to which fish finding, attracting, guiding, surrounding and capturing would be carried out by units remote from the fishing vessel, directed from the vessel, and with little use of human labour. Presumably technical difficulties in building such gears could eventually be solved. Whether they will be built, and then put into commercial use, undoubtedly will depend upon the advantages they might give over present day equipment, and on their cost and efficiency.

No specific conclusion could be reached at this time about the likely efficiency of gears such as those illustrated in earlier paragraphs. One can only say that they would be built only if they could be used to find and capture fish with more certainty than present day gears. The general implication of the illustration is that these gears as both fishing and killing instruments would be constantly nearer to the fish than present-day gears. That is, the seeing and fish-controlling gear would be nearer the fish the echo-sounders and fish finders on surface vessels of today are, in this view, intermediate steps.

In this sense the "fishermen" would see his shoals and round them up almost at will. Some of the uncertainty in present-day fishing would be reduced.

Such equipment, if it came into use, would undoubtedly have considerably greater fishing power. If we could build units which, manned or unmanned, could go below the surface and successfully look for, control and capture fish, operating directly against seen fish instead of simply cast in an area in which fish are known to be, we should also be able to take such fish in large quantities– provided large quantities are present.

BIBLIOGRAPHY

Biswas, K.P., Preliminary observation on the use of electrical seine on the clearance of nursery tanks, *Orissa Fishery Bulletin*, 1, 24-32, 1961.

Biswas, K.P., Preliminary observations on the effect on electrical seine of fish. *Catch Fishery Technology*, 7 (2), 198-206, 1970.

Biswas, K.P., *Ecological and Fisheries Development in Wetlands*, Daya Publishing House, New Delhi, 1995.

Burnet, A.M.R., Electric fishing with pulsating direct current, *New Zealand J. Sci*, 2(1), 46-56, 1959.

Gāmo M. *et al.*, General Research Report on cultural assets of Izu Islands, *Fishery Journal*, 24, 1978.

Gema, G. *et al.*, Sardine fishery by purse seine in the Seto Inland sea: Past and present. *Fishery Journal*, 20, 1981.

Haskell, D.C. *et al.*, Reactions and motion of fish in a direct current electric field, *New York Fish & Game J.*, 1(1), 47-64.

Hiramoto, T., Set net fishery in Kanagawa prefecture, *Fishery Journal*, 22, 1978.

Hirasawal, Y. and Ninohei, T., Mechanization and the family revolution, *Fishery Journal*, 44, 1995.

Holzer, W., Uber eine absolute Reizspannung bei Fishchen (on the absolute stimulus voltage for fish), *Pfluger's Archiv fur die Ges. Physiol*, 229 (2), 153-172.

Inoue, M., Relation of fish movement to fishing methods, *Fishery Journal*, 26, 1979.

Kawasaki, T., Resources of surface fish, *Fishery Journal*, 26, 1979.

Kondo, K., Sardine resources and life pattern, *Fishery Journal*, 16, 1978.

Koyama, T., Thesis on fishing methods, Iustitute of Fisheries Engineering Research, 1975.

McLain, A.L. and Nielsen, W.L., Directing the movement of fish with electricity, *U.S. Fish Wildl. Serv. Spec. Sci. Rep. Fish.*, 93, 24 pp, 1953.

Mc Millan, F.O., Electric fish screen, *U.S. Bur. Fish. Bull.* 44, 97-128, 1928.

Masthawee, P., Purse seining method, *Marine Capture Fisheries*, 10 p., 1982.

Masthawee, P., Braiding and mending of fishing nets, *Marine Capture Fisheries*, 18 p., 1982.

Masthawee, P., Trawl net design., *Marine Capture Fisehries*, 16 p, 1982.

Masthawee, P., Study of net making, *Marine Capture Fisheries*, 10 p, 1982.

Meyers, G. F., The desing of an electric shocker boat, Prog. *Fish. Cult.* 13, 229-231. 1951.

Ministry of Agriculture, Forestry and Fishery, Japan, *Japanese Fishing Gear and Method*, Agricultural Forestry Stat. Association, 1976.

Nagano, T., Structure of Drive-in type bottom lift net fishery in Kozushima. *Fishery Journal*, 24, 1978.

Nakatani, R., The average specific electrical resistance of some salmonids. *Univ. Wash. Sch. Fish. Tech. Rep.* 4, 11 p., 1954.

Nomura, M., *Fishing Techniques*, 1975.

Patten, B.G. and Gillaspie, C.C., The Bureau of Commercial Fisheries type IV electrofishing shocker– its characteristics and operation, *U.S. Fish. & Wildlife Serv. Spec. Sci. Rep-Fish No 529*, 15 p., 1966.

Rollefson, M.D., The development and evaluation of interrupted direct current etectrofishing equipment. *Wyo Game & Fish Comm. Coop. Res. Project.* 1, 123 pp., 1958.

Sigler, W.F., *Electricity in fishery research and management*, Utah. Sci. Agri. Experi. Stat., Utah. St. Univ. 30(3), 72-79, 1969.

Taylor, G.N. *et al.*, Galvanotaxic response of fish to pulsating direct current, *J. Wildl. Mgt*. 21(2), 201-213, 1957.

Thomson, D. B., *Fishing methods, equipment and Deck layout of fishing vessel.*, FAO/SCSP, 1978.

Vincent, R., River electrofishing and fish population estimates. *Prog. Fish. Cult.*, 33 (3), 163-169, 1971.

Yamazaki, T., Purse seine net design, *Marine Capture Fisheries*, 14 p., 1982.

Yamazaki, T., Stick-held dip net (lift net), *Marine Capture Fisheries* 26 p., 1982.

INDEX